Water Always Wins

Water
ALWAYS
Wins

THRIVING IN AN AGE
OF DROUGHT AND DELUGE

Erica Gies

The University of Chicago Press

The University of Chicago Press, Chicago 60637
© 2022 by Erica Gies
All rights reserved. No part of this book may be used or reproduced in any
manner whatsoever without written permission, except in the case of brief
quotations in critical articles and reviews. For more information, contact
the University of Chicago Press, 1427 E. 60th St., Chicago, IL 60637.
Published 2022
Printed in the United States of America

31 30 29 28 27 26 25 24 23 22   1 2 3 4 5

ISBN-13: 978-0-226-71960-3 (cloth)
ISBN-13: 978-0-226-71974-0 (e-book)
DOI: https://doi.org/10.7208/chicago/9780226719740.001.0001

Library of Congress Cataloging-in-Publication Data

Names: Gies, Erica, author.
Title: Water always wins : thriving in an age of drought and deluge / Erica Gies.
Description: Chicago : The University of Chicago Press, 2022. | Includes bibliographical
    references.
Identifiers: LCCN 2021050909 | ISBN 9780226719603 (cloth) | ISBN 9780226719740 (ebook)
Subjects: LCSH: Water conservation. | Water resources development. | Freshwater ecology.
Classification: LCC TD388 .G54 2022 | DDC 333.91/16—dc23/eng/20211116
LC record available at https://lccn.loc.gov/2021050909

♾ This paper meets the requirements of
ANSI/NISO Z39.48-1992 (Permanence of Paper).

*For water and all its relations, be
they mineral or microbial, beaver or bipedal.
And for the water detectives—for their curiosity,
vision, and work toward a livable future.*

Contents

Introduction

On a sunny winter day in San Francisco, Joel Pomerantz brakes his bike in Alamo Square Park near that famous spot where Victorian houses, the Painted Ladies, front the city's modern skyline.

"Do you notice anything?" he asks me.

I brake too and look around, flummoxed. I lived in this city for seventeen years and have been to this park countless times. Everything seems ordinary. On the paved path at our feet, Pomerantz points to an oblong puddle, which I would assume was left over from the last sprinkler watering.

"That?!" I ask, incredulous.

"Look closer," he says, pointing to its ring of mossy scum. "That's a sign that this water is nearly always here." This diminutive puddle, which I have likely passed without noticing many times, is actually evidence of natural springs beneath the park that seep continually, he tells me. It's a small sign of water's hidden life, the actions this life-sustaining compound continues to pursue, despite our illusion that we control it. As climate change amplifies floods and droughts, people like Pomerantz are recognizing the importance of such minutiae that highlight water's agency.

In his free time, Pomerantz hunts and maps ghost streams, the creeks and rivers that once snaked across the San Francisco Peninsula before humans filled them with dirt and trash or holstered them into pipes, then erected roads and buildings atop them. Such treatment of waterways has become standard practices in cities, where more than half of us live worldwide. Pomerantz has devoted three decades to exploring the city with water on his mind, making him a kind of water detective. His eyes see what others miss—like this puddle, or certain water-loving plants that are clues to lost creeks. He gestures toward the trees that line the park's edge on Fulton Street. "Willows are like a flag," he says. In fact, the name of this park is actually a plant clue: *álamo* means "poplar" in Spanish, a species related to willows and other streamside trees.

A few blocks away, he checks for traffic, then guides me to a manhole in the middle of residential Eddy Street near busy Divisadero. Cocking our heads, we hear the sound of rushing water. When that sound is constant, he says, especially in the middle of the night, it's a creek imprisoned in a sewer pipe, not somebody flushing.

Later, Pomerantz and I bike to Duboce Triangle, another small park, this one between the Lower Haight and Castro Districts. Duboce lies at a low point of The Wiggle, San Francisco's beloved bike path. Although unmarked for many years, bikers long followed this route, weaving through valleys at the base of hills. A stream, now buried, was the original traveler of The Wiggle, and along its path through Duboce Triangle the city has now built bioswales, vegetated ditches to hold runoff from heavy rains. Although I've biked the route myself frequently, I never knew it was pioneered by a stream. It makes sense, when you think about it. Cyclists, like water, look for the path of least resistance.

Pomerantz—who has published a map of San Francisco's lost waterways on his Seep City website, advised local agencies as a consultant, and leads walking tours to share his hard-won knowledge—is not alone in his obsession. In Brooklyn, urban planner Eymund Diegel has mapped Gowanus Creek's lost watershed. In Victoria, British Columbia, artist, poet, and environmental activist Dorothy Field worked with local historians and First Nations to track the hidden path of Rock Bay Creek, then installed signs and street medians inlaid with salmon mosaics to draw attention to where it still flows underground. As curiosity about buried waterways grows in the popular imagination, the quirky passion is now a global phenomenon. Subterranean explorers, featured in a 2012 film called *Lost Rivers*, are discovering buried waterways encased in pipes below Toronto, Montreal, and Brescia, Italy. The Museum of London had a *Secret Rivers* exhibition in 2019 to reacquaint Londoners with their lost streams.

Secret rivers, ghost streams, hidden creeks: learning of their existence arouses our innate attraction to mystery and our passion about the places we live. What we learn about the past triggers amazement because our quotidian landscape is so transformed. We've dramatically altered waterways outside of cities too. We've straightened rivers' meanders for shipping, uncurled creeks to speed water away, drained and filled wetlands and lakes, and blocked off floodplains to create more farmland or real estate for buildings.

But our curiosity about water's true nature is not idle, nor an indulgent wish to return to the past. Water *seems* malleable, cooperative, willing to flow where we direct it. But as our development expands and as the climate changes, water is increasingly swamping cities or dropping to unreachable depths below farms, generally making life—ours and other species'—precarious. Signs of water's persistence abound if we train ourselves to notice them. Supposedly vanquished waterways pop up stubbornly, in inconvenient ways. In Toronto, tilted houses on Shaw Street near the Christie Pits neighborhood were long a local novelty, but most people didn't know that the ghost of Garrison Creek was pulling them out of plumb. Worldwide, seasonal creeks emerging in basements are evidence that those houses encroach on buried streams. In my partner's mom's neighborhood in suburban Boston, most houses come with sump pumps because the development was built on the local "Great Swamp." And in the wreckage of disasters like Superstorm Sandy or Hurricane Harvey, we see that homes built atop wetlands are the first to flood.

When our attempts to control water fail, we are reminded that water has its own agenda, a life of its own. Water finds its chosen path through a landscape, molding it and being directed in turn. It has relationships with rocks and soil, plants and animals, from microbes to mammals like beavers and humans. Today, water is revealing its true nature increasingly often, as climate change brings more frequent and severe droughts and floods. To reduce the impacts of these phenomena, water detectives—Pomerantz and other ghost-stream enthusiasts, restoration ecologists, hydrogeologists, biologists, anthropologists, urban planners, landscape architects, and engineers—are now asking a critical question: What does water want?

California: Where Water Fixations Are Born

Figuring out what water wants—and accommodating its desires within our human landscapes—is now a crucial survival strategy. My own preoccupation with water goes back to childhood. I grew up in California, where jockeying over this molecule has been the unofficial sport since before the state's founding. Water first grabbed my attention during the 1976–77 drought, when grade school assemblies taught us to take quick showers rather than deep baths, and my dad put a brick in our toilet tank to displace volume and reduce the amount we flushed. That message of scarcity, that this vital resource is precious, embedded deeply in my brain.

I also dipped into water's wilder side. My family frequently went camping with another, who had a daughter my age. As California girls, raised on the beach and among the redwoods, we took it as a point of pride to go swimming in any body of water we came across, from big breakers off Santa Cruz to eight-thousand-foot-high alpine lakes partially covered in ice. So it's small surprise that, when I became a journalist, I was drawn to covering California's incessant water wrangles.

The first thing I came to understand—which sounds obvious but I hadn't before given it much thought—is that much of the water we see today, especially in industrialized countries, is not in its natural state. Humans' efforts to control it have created giant lakes behind dams; deeply scoured, fast-flowing rivers; straightened, narrow creeks far below their banks; arrow-straight canals that deliver irrigation water to farms. We've also erased many lakes and swamps entirely. Worldwide, only one-third of rivers longer than 620 miles travel uninterrupted to the ocean. Most of the remaining free rivers run through remote parts of the Arctic, Amazon, and Congo. All the rest have been dammed, straitjacketed in levees, and dredged to make shipping channels.

What many of us think of as "river" is a hobbled water canal that no longer wanders across its floodplains, depositing nutrient-rich, land-forming silt as it goes. As a kid, the wildest river I knew was the Sacramento. Although nearly a mile wide at its mouth near the East Bay town of Antioch, the Sacramento and its major tributaries are largely tamed by giant dams in the north—Shasta, Oroville, Canyon, Folsom—and the main stem is heavily constrained as it flows more than four hundred miles to the sea. Even in its fan-shaped delta, fingers of water are hemmed in by dirt levees built by hand a century ago. Today the farmland inside those cordons has sunk up to twenty-six feet from the loss of peat, which decays as it dries out, and from blockage of the river's natural deliveries of sediment that continually build land.

The first time I saw a truly wild river was on a visit to Alaska's Denali National Park. In late August, I was hiking with friends through spongy tundra, its bonsai shrubs on fire with autumn colors. We chatted loudly as we went to avoid surprising grizzly bears, until we came to the McKinley River. The water spanned maybe forty feet across. Out for a day hike, unprepared to take it on, we had to turn back. As I gazed up the valley from which it flowed, I saw that the barrier blocking us was just one strand of a broad, braided system spread languidly across the floodplain, its columns shifting apart, then twining together. Although at that moment I knew little about hydrology, the science of water, on some instinctual level I understood that this was a free river. And every other river I'd known was markedly subdued.

Letting Go of the Control of Nature

Humans have sought to control our environment throughout history. In looking for an antonym for *control*, I found *chaos, lawlessness, mismanagement, neglect, weakness, powerlessness, helplessness*. It's a linguistic reflection of how much we crave control and fear letting go. But if we want to solve the water problems we face today, we need to open our minds. The way we relate to water is not inevitable. And in fact, our infrastructure, our laws for allocation, our striving for control are amplifying these problems. By asking, "What does water want?" water detectives are working from a philosophy rooted in curiosity, respect, and humility, rather than a too-common arrogance. They are also accepting reality: water always wins.

Certainly that's true in geologic time. If water were a category in the game rock, paper, scissors, water would beat them all every time. Part of the sense of awe we get at the Grand Canyon is wrapping our minds around the fact that the reflective squiggle a mile below us carved that natural cathedral out of rock over millions of years. But water also wins by breaking through our dams and levees sooner or later—in a few months, years, or decades. Today's water detectives are acknowledging water's power and aspiring to go with the flow rather than fight it.

The detectives start by uncovering what water did before generations of humans so radically transformed our landscape and waterways. How did water interact with local rocks and soils, ecosystems and climates before we scrambled them?

There are many ways to get to know water's habits and relationships with other entities. One of them is Pomerantz's approach of close observation. Other water detectives we will meet in this book use magnetic imaging, satellite data, chemical analysis, soil core samples, anthropological research, biology, ecology, and more to ferret out what water is doing. Historical ecologists pursue this knowledge by performing a kind of forensic ecology. As the detectives seek answers to water's true nature and make intriguing discoveries, we begin to understand why certain areas flood repeatedly, or how our tendency to speed water off the land deprives us of urgently needed local rainfall. Then we begin to think creatively about how we can solve these problems by making space for water within our existing habitat.

Some of the detectives' early insights astound—and should spur us to change our ways. For example, the carbon dioxide stored in some wetlands is vastly greater than that stored in forests. Surface water and groundwater are not separate sources but part of a single, interactive system. When rivers shrink to a trickle, water underground can feed them, pushing up through the bottom of the streambed. Conversely, when people pump groundwater and the water table drops, river water can filter through the bottom of the streambed to replenish it. Tidal marshes can actually keep pace with sea-level rise, protecting inland areas, if they have enough sediment—a resource we've made scarce with our widespread dam building and river channelization.

The answers the water detectives are discovering in the cities, fields, swamps, marshes, floodplains, mountains, and forests that we will journey through in this book lie in conserving or repairing natural systems, or mimicking nature to restore some natural functions—not building more concrete infrastructure. These ecosystems can buffer us from bigger rainstorms and longer droughts by absorbing and holding water. When we obliterate them, we make our places brittle, multiplying the intensity of these disasters.

Among water professionals around the world, these reparative approaches go by various names, including nature-based systems or solutions, green or natural infrastructure, sponge cities, low-impact development, and water-sensitive urban design. Because these solutions seek to work with or simulate natural systems, they offer myriad benefits beyond just reducing floods and droughts. For example, they help us address another threat to life as we know it: the dramatic decline of other species that we are causing. Also, because natural systems store carbon dioxide in plants and soil, they help us not just adapt to climate change but also slow its progression. Protecting biodiversity and storing carbon are not peripheral to solving water problems; they are integral to healthy water systems.

Slow Water Manifesto

So what *does* water want? Most modern humans have forgotten that water's true nature is to flex with the rhythms of the earth, expanding and retreating in an eternal dance upon the land. In its liquid state, with sufficient quantity or gravity, water can rush across the land in torrential rivers or tumble in awe-inspiring waterfalls. But it is also inclined to linger to a degree that would shock most of us because our conventional infrastructure has erased so many of its slow phases, instead confining water and speeding it away. Slow stages are particularly prone to our disturbance because they tend to be in the flatter places—once floodplains and wetlands—where we are attracted to settle.

But when water stalls on the land, that's when the magic happens, cycling water underground and providing habitat and food for many forms of life, including us. The key to greater resilience, say the water detectives, is to find ways to let water be water, to reclaim space for it to interact with the land. The innovative water management projects I visited around the world all aim to slow water on land in some approximation of natural patterns. For that reason, I've come to think of this movement as "Slow Water."

Like the Slow Food movement founded in Italy in the late twentieth century in opposition to fast food and all its ills, Slow Water approaches are bespoke: they work with local landscapes, climates, and cultures rather than try to control or change them. Slow Food aims to preserve local food cultures and to draw people's attention to where their food comes from and how its production affects people and the environment. Similarly, Slow Water seeks to call out the ways in which speeding water off the land causes problems. Its goal is to restore natural slow phases to support local availability, flood control, carbon storage, and myriad forms of life. For many people who study water deeply, these values have become obvious.

Just as Slow Food is local, supporting local farmers and thereby protecting a region's rural land from industrial development and reducing food's shipping miles and carbon footprint, ideally, Slow Water is too. The engineered response to water scarcity has been to bring in more water from somewhere else. But desalinating water or transporting it long distances consumes a lot of energy: in California, for example, the giant pumps that push water southward from the Sacramento Delta are the state's largest user of electricity. Withdrawing water from one basin and moving it to another can also deplete the donor ecosystem, or introduce invasive species to the receiver ecosystem.

Perhaps the biggest problem with bringing in water from somewhere else is that it imparts a false sense of security. When we live long distances from our water, we don't understand the limits of that supply, so we're less likely to conserve. We also don't understand how the water we use supports its local ecosystem. By overexpanding human population and activities, especially where there isn't enough local water, such as in the US Southwest, Southern California, or the Middle East, we make people and activities vulnerable to the water cycle, rather than resilient.

Slow Water is also in the spirit of the land ethic articulated by twentieth century forester-turned-conservationist Aldo Leopold. It calls for us to treat soil, water, plants, and animals with respect and to strengthen our relationship with them because they are part of our communities and we have a moral responsibility to them. His hydrologist son, Luna Leopold, expanded these ideas into a water ethic that calls for "a reverence for rivers." Both ethics express an interweaving of nurture and need: for nature to provide for us, we must care for it.

Aldo Leopold was inspired by older traditions. Kelsey Leonard is a Shinnecock citizen and assistant professor in the School of Environment, Resources, and Sustainability at the University of Waterloo in Ontario. As she explained to me and an audience of river researchers in an online talk in 2020, many Indigenous traditions don't consider water to be a "what"—a commodity—but a "who." Many Indigenous people not only believe that water is alive, but that it's kin. "That type of orientation transforms the way in which we make decisions about how we might protect water," she said. "Protect it in the way that you would protect your grandmother, your mother, your sister, your aunties."

Such belief that natural things are alive, or have souls, including rivers, rocks, trees, animals—often called animism—is common in ancient thinking worldwide. Similar beliefs elsewhere include Bon, the precursor to Tibetan Buddhism, and Celtic and Norse beliefs in fairies and elves, the spirits of the grasslands and forest, still held today by many people. From this world view comes the Indigenous water protectors' rallying cry, "Water is life."

In contrast, today's dominant culture is rooted in an ideology of human supremacy: humans' needs and wants—particularly privileged humans—are considered more important than other species' right to exist. (The attitude of supremacy extends to "othering" certain people too.) This us-first stance hasn't done humanity any favors. By focusing single-mindedly on servicing human needs, we ignore other interconnected entities in the systems we change, causing myriad unintended consequences, from climate change to the extinction of other species to water woes. It's also a moral issue, as the Leopolds and Leonard point out: humans are not, in fact, more important than other beings. They, like us, have a right to exist.

Leonard says that one way to solve many water injustices is to recognize water as a legal person with an inherent right to exist, flourish, and naturally evolve. That's not as radical a notion as it might sound: in the United States, corporations were granted legal personhood with all the rights that implies. According to Indigenous beliefs, water is actually alive, while corporations are not. "Who is justice for? Humanity alone?" Leonard asks rhetorically. Indigenous legal systems already protect nonhuman relations, including water.

In fact, a rights of nature movement is starting to infiltrate Europe-based legal systems. Dating back to the 1970s, it argues that nature has a fundamental right to exist. A Pennsylvania-based legal advocacy organization, the Community for Environmental Defense Legal Fund, uses this argument to assert a community's right to prevent a corporation from polluting its territory. Ecuador and Bolivia, which have large Indigenous populations, have enshrined rights of nature in their constitutions. In New Zealand, Whanganui River, sacred to the Indigenous Māori people, has won legal personhood. Same with the storied Ganges River in India and the Magpie River in Quebec that is sacred to Innu people. The Yurok Tribe in Northern California has granted legal personhood to the Klamath River. Other communities around the world are also fighting for legal rights for their rivers, wetlands, and watersheds. A river's rights can include the right to flow, the right for its cycles to be respected, the right for its natural evolution to be protected, the right to be free from pollution, the right to maintain its natural biodiversity, the right to fulfill its ecosystem's essential functions. With legal personhood, if these rights are violated, people have the standing to sue on the river's behalf.

The water detectives in this book are a diverse bunch and don't all hold these beliefs. But they share an openness to moving from a control mindset to one of respect. Their openness is at the heart of this book. As our long-held illusion that we can control water is crumbling in the face of escalating disasters, we understand, viscerally, that water always wins. Given that truth, it's better to learn how to accommodate water, to work with water, and enjoy the benefits that cooperation can bring.

While politics and finance are fundamental to getting anything done in this world, they are not the primary focus of this book. Similarly, although agriculture is the major human use of water worldwide, I do not delve deeply into the industrial water-energy-food nexus, which would be a tome in its own right. Instead, this book aspires to spark curiosity about what water wants by looking at its physical relationships with other entities.

Within this book I follow a loosely chronological order. Chapter 1 lays out the problems we're currently creating with water. The rest of the book introduces Slow Water approaches from around the world, organized around water's relationship with different natural elements through time. Chapter 2 focuses on water in rock and soil underground: geologic time. Chapter 3 looks at water's interaction with early life: microbes and slightly bigger creatures, meiofauna and macroinvertebrates. Chapter 4 attends to bigger critters—especially that renowned furry water engineer, the beaver. Chapters 5 and 6 look at humans reviving ancient techniques that work with nature to manage water. Chapter 7 addresses the Industrial Era: how did mainstream culture's attitude toward water change? Chapters 8, 9, and 10 provide glimpses of the near future and people's adaptations: natural water towers, coastal restorations—and retreat.

Slow Water solutions are gaining momentum worldwide, so the places we visit in this book—the United States, Canada, Iraq, the United Kingdom, India, Peru, China, the Netherlands, Kenya, Vietnam—are by no means comprehensive. Rather, they represent a range of continents, peoples, ecosystems, and water problems to show how each place is unique—yet all share common concerns. In these chapters, people are grappling with droughts and floods, melting glaciers and reduced monsoons, sinking ground, soil erosion, decline of other species, sea-level rise, and salt water moving inland. The places featured don't necessarily have it all figured out. Some bold ideas are being implemented piecemeal, as people struggle to get them off the ground or expand to scale. Some governments have policies contradictory to their visionary Slow Water approaches. Healing our relationship with water is a process. In spending time with the water detectives, I've learned a lot about what water wants and how it has shaped and been shaped by various entities through time. My hope is that these stories of innovation can inspire us to think differently about water and the ecosystems we share so we can harvest ideas to try in our own places.

· 1 ·

Descending into Chaos

The speed of the water is what shocked many New Yorkers. On October 29, 2012, as Superstorm Sandy was homing in on the US East Coast, John Cori decided to sit tight. A lifetime resident of the Rockaway district of Queens, he had ridden out plenty of tempests in his house, located about seven hundred feet from shore. But then the storm hit, slamming into the coast.

"The water was rolling down the streets like a river, getting bigger and bigger and bigger," Cori tells me over the phone, recalling the surreal experience. Over a twenty-minute period, he watched as it reared up from one foot deep to five, blowing apart his neighbor's fence. "It looked like Niagara Falls, coming from the ocean." He began to question the wisdom of staying put when part of the Rockaway Beach boardwalk floated like a giant barge down the street and crashed into his porch. Then the water worked the boardwalk loose, pushing it into his neighbors' house, wedging it there: "That was the scary part because it was acting like a dam." Cars, sand, and other debris backed up in front of it. It was so astonishing that it reminded him of another unbelievable event. "To me, it was like looking at the World Trade Center coming down," he explains, in the sense that "I could not believe this is happening."

It may sound like sacrilege to compare a storm to 9/11. But Cori isn't the only New Yorker to draw that analogy. I visited the city in early 2019 on a journalism fellowship focused on resilience, and person after person recounted the trauma of Sandy. On another follow-up phone call, I talk to Captain Jonathan Boulware, executive director of Lower Manhattan's South Street Seaport Museum, which tells the history of New York as a port city. "It was a completely different kind of event than a terrorist attack, but there are similarities," he tells me. "I know exactly where I was for Sandy. I know where I was for 9/11. It's one of those events in my life. And it forever altered the museum, the city, and me." When Sandy was over, at least 233 people were dead, and economic losses totaled more than $60 billion.

Boulware trained as a sea captain, and it shows in his calibrated, hypercompetent demeanor and deep knowledge of the ocean. He was at the museum when Sandy struck. "I was in the lobby and I could hear this terrible sound, like a cataract of water. And it was coming through vents in the floor." This confused Boulware because outside, the streets were still dry. Although you might expect that a hurricane would push surging waves across the land, in Lower Manhattan, basements first flooded from below. Rising water from the storm surge, the East River's flow, and a high tide amplified by the moon put pressure on groundwater, pushing it up through the subway system and into buildings' interconnected stormwater and sewer drains. Sandy's stormfront spanned a thousand miles, more than three times the size of Hurricane Katrina, the epic storm that swamped New Orleans in 2005. The size increases the height of the storm surge, which for Sandy was about fourteen feet, blowing away the previous record of ten feet during 1960's Hurricane Donna. As Boulware and a colleague watched from an upper floor, the water rose in the lobby, higher and higher, to more than six feet deep, overtaking the museum's front entrance. By then water was also flowing through the streets outside.

"Were you afraid?" I ask.

He inhales. There is a long pause. "It didn't even occur to me to ask myself that," he says finally. As a sailor, watching water flood in where it shouldn't be triggered a reflexive response: prepare to abandon ship. "And then I thought, that's an idiotic idea. Where are we gonna go? We're in a building that is surrounded by five, six, seven feet of oily, toxic, fast-moving water." He reflects. "Yeah, I think there was some fear. We were immobile, we were trapped, and there wasn't anything we could do about it."

Finally, Boulware caught a few hours' sleep, awaking before dawn to find the water gone. He stepped out onto streets covered in a slick of oil and broken glass. Shop windows had burst outward from the pressure of interior floods. "The buildings had sort of vomited their contents. . . . That stuff was all over the street."

Further inland, across the Hudson River in Hoboken, New Jersey, my friend and fellow environmental journalist Sharon Guynup tracked the storm from her apartment in the middle of town, about a mile from the river. At 9 p.m. she went outside to take a look around. "A block and a half away, I saw a wall of angry water, three to four feet deep, coming raging up the street," she recalls. "It looked like a disaster movie." Shock turned to sheer adrenaline when she turned and saw another surge coming from the west, having gushed across town and circled back. The two waves were so powerful that they blew the metal basement door off its hinges, destroying everything her family had in storage.

The next morning, her building and much of the city were sitting in a giant lake. Because her apartment was on the second floor, she was a foot clear of the water. She and her family camped there with no running water or electricity for four days, subsisting on rationed water and cooking on their gas stove as the water receded. Then the temperature dropped to freezing, and they all had splitting headaches from breathing intense fumes from the polluted water. Guynup, her husband, and her adult son crammed into a car with their dog, puppy, and cat and drove around for an entire day, searching for a hotel with electricity and space. The single room they finally found was home for the next three weeks.

Guynup had been writing about climate change for fifteen years. "But with Sandy, climate change got personal. It's very, very different when it happens to you. Now, whenever I see flooding"—today, virtually a constant somewhere in the world—"I feel it in *my gut*," she says, stressing the last syllables. "I feel the devastation. That's a trauma that I carry with me."

*

At the same time, across the country, California was in the opening months of a much slower-moving disaster: a brutal drought that would drag on for five years—with effects that continue to play out in the state's apocalyptic firestorms and ongoing water scarcity. Stagnant pools languished at the bottom of reservoirs; boats came to rest on the desiccated mud of receding shorelines; lawns turned crunchy (some with signs extolling water conservation: "Brown is the new green"); pine forests died, their orange needles a visual alarm. As the months and then years ticked by, 100 percent of the state was in drought. On a summer road trip down the Redwood Highway in the typically damp northwestern corner of the state, I was shocked at the condition of the two-hundred-mile-long Eel River that drains five counties. Beneath the towering trees, it was reduced to a brown trickle. Campers sat on lawn chairs right in the riverbed to get close to its remains. All these signposts created a subconscious drumbeat: scarcity, scarcity, scarcity. The fear was palpable: Can we continue to live here? Can we support more new residents? Should we be growing food to export to the world when agriculture takes 80 percent of the state's human-used water but crops provide only around 1.4 percent of its total economy?

Aside from the cultural anxiety of water scarcity, the drought prompted real-world reductions. In 2015, then-Governor Jerry Brown required water supply agencies throughout the state to reduce consumption by 25 percent, and more cities turned to water reuse or dusted off plans for desalination plants. But these moves can seem a bit peripheral given agriculture's outsize water footprint. (California is not an outlier; worldwide, agriculture accounts for 70 percent of human water use.) Because state water rights go first to people with the oldest claims, in 2015 farmers with "junior" water rights saw their allocations cut to zero, and more than a million acres of Central Valley cropland was fallowed—about 2.5 times more than in 2011, according to satellite-based estimates by NASA researchers. For land that remained under cultivation, much of the shortfall was covered by farmers pumping unregulated groundwater. As a result, between 2011 and 2015, California lost more than five trillion gallons of water from its underground aquifers, part of a decades-long trend. Now that long-term groundwater borrowing has come due: in a 2019 report, the Public Policy Institute of California determined that 535,000 acres in Southern California's dry San Joaquin Valley must be fallowed permanently by 2040 to allow emptied aquifers to recover.

California has had six major droughts since record keeping began, and three have been in the twenty-first century. In 2020, it was entering another one. Climate change is making dry periods more intense by causing a "thirstier" atmosphere that evaporates more water from soil and reservoirs. The 1976–77 drought that instilled my water consciousness was more severe than the one in the 2010s, but also shorter, so its impact on water supplies was smaller.

In late 2016, just as the thirst was growing truly desperate, the atmospheric rivers arrived. These enormous trains of condensed vapor move from west to east across the Pacific like rivers in the sky, 1,200 to 6,000 miles long and capable of carrying fifteen times the average flow at the mouth of the Mississippi. When they hit mountains in California, they can dump hurricane quantities of rain. Storms borne on atmospheric rivers are not new in the state, but they are increasing in frequency worldwide, likely doubling by the end of this century. And as Earth continues to warm, models show that atmospheric rivers in California will hold an estimated 10 to 40 percent more water, increasing the risk of floods and mudslides.

Fifteen of these strong atmospheric rivers hit California during the winter of 2016–17, turning hillsides so sodden that they slid down over roads. "Seventeen is closed, Skyline is closed, 9 is closed, 152 is closed," my mom told me that February, rattling off highways between the San Francisco Bay Area and the Pacific coast, where I was scheduled to drive the next morning. The mudslides cut off communities along Big Sur's rocky cliffs for more than a year. North of Sacramento, more than 188,000 people were forced to evacuate their homes as overflow at the Oroville Dam eroded a giant crater into its spillway, threatening a massive flood by releasing one of the state's biggest reservoirs all at once. Although disaster was averted, it cost $1.1 billion to fix. After the years of desperate want, the sudden rush of water was a mind-bender. At first, the deluges brought relief, painting chaparral hills an uncharacteristic fresh green and spawning super blooms across wildlands. But that feeling shifted to growing unease, as reservoirs spilled over into swollen creeks. "I can't remember ever seeing it like this," my mom says, echoing the amazement of millions of longtime Californians.

*

More and more people around the world are facing such difficult, expensive disasters as climate change ramps up. Global economic losses from flooding rose from $500 million annually, on average, in the 1980s to $76 billion in 2020. The World Meteorological Organization expects the number of people at risk from flooding to increase by nearly a half billion by 2050. When it comes to drought, already more than two billion people around the world live with severe or high water insecurity. As the climate continues to warm, two-thirds of the global population is predicted to experience a progressive increase in drought conditions. We are also making the problems worse with our development decisions. For example, New York has filled in and hardened around 85 percent of its coastal wetlands, removing their ability to absorb water and building new developments in harm's way. And California's current drought is made more desperate because of vast demand for water from the ever-expanding population and mega food production system that's propped up on supply from the northern third of the state and elsewhere.

Water is complicated, so solving these problems will be too; yet it's also fascinating. And it's not too late to turn things around. But we need to think and act differently. Instead of more engineering, nature can be our buffer. "Natural ecosystems are humanity's first line of defense against floods, droughts, hurricanes, heat waves and the other mounting impacts of climate change," according to a 2019 report from the Global Commission on Adaptation. That's because forests and wetlands evolved to withstand a wide range of natural variation in water supply. While a levee, for example, confines water within a narrowed river channel, raising its level and pushing the problem downstream, a floodplain accommodates what water wants: to spread out, slow down, and rejoin the river and groundwater in due time. Returning space to water near our homes and businesses creates more pliant boundaries to absorb floods and to keep the local water supply robust. Water fluctuations are just part of natural cycles, not disasters, if human activities aren't devastated.

The complexity of natural systems also offers myriad benefits compared with, say, a dam built for one purpose: to hold water or generate electricity. Wetlands, floodplains, and forests don't just mitigate floods and droughts. They are home to other animals and plants, store carbon dioxide, anchor topsoil, reduce the need for irrigation and fertilizer, clean water, build land, and give depressed urbanites a needed dose of nature. Even better, healthy ecosystems—if we give them space to do their thing—maintain themselves. That's not true of concrete interventions, which is a reason nature-based solutions are often cheaper.

Climate Crisis = Water Crisis

While floods and droughts occur naturally, they are becoming more frequent and intense because of climate change. In the United States, the amount of precipitation falling during heavy storms has increased dramatically since 1958: by 55 percent in the Northeast, 27 percent in the Southeast, 42 percent in the Midwest, 10 percent in the Southwest, and 9 percent in the Northwest. For example, Hurricane Harvey dumped fifty inches of rain on Houston over four days in 2017, causing the area's third five-hundred-year flood in three years. Scientists at Lawrence Berkeley National Lab calculated that climate change increased precipitation during Harvey by as much as 38 percent. That same year, Hurricane Maria dropped forty-one inches of rain on Puerto Rico, more than any storm since 1956. Researchers concluded that a storm of Maria's magnitude is now five times more likely to form than in 1950 due to climate change.

Conversely, the western United States is suffering a climate-driven megadrought, going on twenty years and counting. (Researchers disagree on whether California is also suffering from this megadrought, the big storms notwithstanding.) On the other side of the globe, the worst drought in five hundred years was one factor in touching off Syria's civil war, which has killed about 585,000 people and displaced twelve million. In the Himalayan Mountains, glaciers that provide a steady water supply for many millions of people downstream by melting slowly through the summer are disappearing. On our current greenhouse gas emissions trajectory, two-thirds could be gone by 2100.

There are several ways in which climate chaos scrambles the water patterns we've relied on over the past couple hundred years. As we've come to understand, water moves across Earth's surface, into the atmosphere, and underground as liquid, ice, or vapor. Heat from the sun causes liquid water—in lakes, rivers, and oceans—to turn into gas and evaporate into the air. Water vapor condenses into clouds, cools, then falls as rain or snow. Snow and ice accumulate and then melt, feeding streams, rivers, and lakes. Plants and soil absorb water and release it back into the air as vapor via evapotranspiration, like an exhalation. Some surface water seeps into the ground, feeding groundwater such as aquifers and springs, or travels deep into Earth until it is returned to the surface via volcanic steam or geyser. This is the water cycle.

Over the last century, the pace at which humans emit greenhouse gases such as carbon dioxide and methane has been accelerating dramatically. Just in my lifetime, people have released more emissions than during the ten thousand years of human civilization that came before, increasing the average temperature globally by two degrees Fahrenheit since 1880. Oceans have absorbed more than 90 percent of the excess heat since the 1970s. Warmer seawater, especially in tropical areas, evaporates more readily, increasing air temperature and humidity, leading to increased rain in some areas.

Warmer air also evaporates more water out of soil and freshwater bodies and allows the atmosphere to hold more water. That's be-cause the molecules of vapor are moving faster than those in colder air, making them less likely to condense back into liquid to fall as rain. For each Celsius-degree increase of warming, the air holds about 7 percent more vapor. And because water vapor is itself a greenhouse gas, more water in the air creates a feedback loop, fur-ther warming the globe and accelerating climate change. But while some places are getting more rain, others are experiencing increased water scarcity. Warmer temperatures are also drying out the land through evapotranspiration.

Water expands as it heats up, so warming ocean temperatures are increasing the space that seawater occupies. That plus melting ice are raising sea levels and altering ocean currents, which play a role in changing weather patterns. For example, the faltering Gulf Stream current in the Atlantic Ocean is a contributing factor in Europe's hotter summers and colder winters. Higher, warmer oceans mean bigger storm surges, as we saw with Sandy. Global mean sea level has risen eight to nine inches since 1880 and is accelerating. Places such as Miami and Norfolk, Virginia, are among the cities that now flood frequently at high tide, and many places along the US coastline flood three to nine times more often today than fifty years ago. If we don't make significant changes to slow climate change, seas could rise more than eight feet by 2100. More than 40 percent of people worldwide live in coastal areas, where they are vulnerable to flood-ing. People living upriver from coasts are also experiencing more frequent flooding as higher sea levels push water inland.

Weather catastrophes intensified by the climate crisis are now happening about once a week somewhere in the world, causing death, suffering, and displacement, the UN said in 2019, although most of these events fly under the radar of our attention.

Land Follies

Climate change is just one factor, though, in the increasing fre-quency and severity of flood and drought. These events are made more disastrous because of the way mainstream society has developed, and not just by building in vulnerable locations—like in California's semi-arid climate or close to sea level in New York. We are also destroying wildlands and the plants and animals who live there and keep these ecosystems functioning. In our efforts to control water, we are ignoring its relationships with these entities. Slow Water solutions have the potential to address these problems at multiple scales.

We have significantly altered 75 percent of the world's land area for such uses as housing, agriculture, and industry. And a recent study using different criteria found less than 3 percent of land is considered ecologically intact. It's not a coincidence that these rates have soared as the global population has nearly quadrupled since 1900, from fewer than two billion people to more than 7.9 billion. And while the rate of population growth is slowing, UN demographers say the planet will hit around 10.9 billion humans by 2100. (The well-established path to slowing population growth includes access to birth control and education for women and making it acceptable—in all cultures—to have one or zero kids.)

Consumption is another factor in environmental degradation. For example, many of us consume wasteful amounts of water, especially when including the quantity needed to produce some of the goods we buy. A cup of coffee requires more than thirty-four gallons of water; a nine-ounce sirloin steak requires about 1,018 gallons. If every person on the planet enjoyed the average American's level of resource consumption, some researchers estimate Earth could sustainably support about two billion people. If left unchecked, by 2050, human resource use may more than double over 2015 levels; however, many environmental limits are already maxed out globally. But human existence doesn't have to equal land degradation. On the more than 25 percent of the world's land that is managed by Indigenous peoples, nature is healthier and biodiversity is higher. This fact is one of the arguments made by the Indigenous Land Back movement, which advocates that public land, for starters, be returned to Indigenous peoples.

The rapidly sprawling cities we build to serve our population and consumption exacerbate both floods and droughts. As Joel Pomerantz in San Francisco and other ghost-stream hunters are highlighting, humans have constrained space for streams and rivers. Since the 1700s, we've filled or drained as much as 87 percent of the world's wetlands that stored water, held carbon dioxide, and housed countless critters. We've packed them with dirt and poured asphalt on top. These actions also exacerbate local water scarcity, as impervious surfaces—buildings, streets, parking lots—block rain from absorbing into the ground.

Rural lands are also suffering at our hand. Converting wilderness into conventional cropland dries out the land by denuding it of the plants whose roots hold water in the soil and whose leaves release water into the air. Their removal also decreases future rainfall. Plowing and livestock cultivation dry out the soil too, and compact it, collapsing pockets for air and water, making the land less able to absorb rains when they come. Along coasts, pumping groundwater can create a vacuum that pulls seawater inland underground, turning crop land saline, a phenomenon happening now in many places, including Vietnam and Oceanic nations such as Kiribati, Tuvalu, and the Marshall Islands.

Life Loss

Our resource consumption also has an inverse relationship with biodiversity—the richness of life forms in ecosystems. It's understandable: that 75 percent of land area we've impacted was previously home to many plants and other animals. Since 1970, as our global population has more than doubled, the average population size of wild animals has declined by more than two-thirds globally. Today, approximately 25 percent of known species worldwide—about one million—face extinction at our hands, a rate up to a thousand times faster than what's natural, according to a UN-commissioned assessment.

These losses are all the more tragic because we are not just meeting our needs but fueling excessive materialism to further enrich a tiny minority of humans. The price of generating this wealth is paid by everyone else—especially marginalized communities and other species. It also threatens human survival. Interwoven life processes create many of the things we need. For example, pollinators help us grow food, microorganisms and insects break down and recycle waste, predators keep prey in check so they don't denude the land. These systems also provide critical water services such as cleaning, supply, and flood protection. The world is trying to protect biodiversity via an international compact similar to the Paris Agreement on climate change. Called the Aichi Targets, the compact set goals for 2020, including protecting 17 percent of countries' land and inland seas for other species. World governments mostly failed to meet those targets but are expected to set a new goal to protect 30 percent of their land and oceans by 2030. American biologist E. O. Wilson thinks we need to go further to save ourselves and is calling for 50 percent of Earth to be protected, a vision outlined in his 2016 book *Half-Earth: Our Planet's Fight for Life*. Which areas are designated and how effective the protections are key questions. A 2020 study found that restoring the right 15 percent of converted lands could prevent 60 percent of predicted extinctions and store nearly three hundred gigatons of carbon dioxide—30 percent of the total we've emitted since the Industrial Revolution began.

The effort to conserve biodiversity stands to gain a lot from joining forces with Slow Water, and vice versa. That's because healthy freshwater ecosystems sustain greater numbers of species per square mile than almost any other type. And these plants and critters have been particularly hard hit by all our mucking around with water. Species living in rivers and lakes are declining more than twice as fast as land-based and ocean species; they've lost more than 81 percent of their populations, on average, between 1970 and 2012. Conversely, lands and waters rich in biodiversity are healthier and do a better job of providing water services and soaking up greenhouse gases. That's why conserving biodiversity is integral to smart water management: plants and other animals living their lives is what allows ecosystems to self-regulate. (This is also a happy synergy for me, personally. I'm an inveterate lover of all critters, so researching Slow Water projects has been a welcome excuse to spend time in their habitats and see them going about their lives.)

Still, the dominant economic system takes these natural systems for granted, assuming they will always be there. People focused on profits tend to argue that protecting the environment will depress economic activity. But as former Wisconsin governor and Earth Day founder Gaylord Nelson famously said, "The economy is a wholly owned subsidiary of the environment." Without the environment, and the resources and services it provides, there would be no economy.

Land Overlooked in Carbon Emissions

The way we alter land also causes climate change directly. People trying to solve climate change often talk about shifting human energy use to renewable sources such as solar and wind and away from fossil fuels, which have played a large role in getting us into this mess. But there's another big source of emissions that is often overlooked: clearing land for agriculture, industry, and housing. Land-use change causes 23 percent of global greenhouse gas emissions, according to a 2019 report from the Intergovernmental Panel on Climate Change (IPCC).

That's because when we cut or burn live plants, such as trees or grasses, we release the carbon dioxide they're storing and stop them from continuing to absorb more. Stirring up soil kills tiny animals and organic matter, releasing their stored carbon as well. Healthy wetlands, tidal marshes, peatlands, and forests, on the other hand, are huge carbon and methane sinks. But if we don't protect them— and permafrost melts, releasing long-stored methane; carbon-rich rainforests transition to savanna and grasslands; boreal forests die— their escaping greenhouse gases could accelerate climate change. In this way, destroying land's natural processes is a double-whammy from a global warming perspective.

On the other hand, nature-based solutions such as conserving or restoring wetlands and forests are a huge opportunity to both dramatically reduce the emissions causing climate change and adapt to what's already happening by creating more flexible, resilient buffers around our human habitat. Nature-based solutions can provide 37 percent of the climate mitigation needed between now and 2030 to keep global warming below two degrees Celsius—the target of the Paris Agreement. But these interventions currently receive just 3 percent of climate funding. Given that climate change could cost the world 11 to 14 percent of global economic output by 2050, investing more now in nature-based solutions would ultimately save money.

Water Follies

Ironically, we're making water extremes worse by trying to solve flooding and drought. For millennia, humans have built infrastructure to optimize access to water, especially with the advent of agriculture. But in the Industrial Age, with our horsepower amplified by fossil fuels and with the exponential population growth in the last couple of centuries, the scale of our interventions, such as dams, levees, and channelized rivers—and their impacts—have grown apace.

Dubbed *gray infrastructure* by planners because they are often built with concrete, the water management systems we've built are causing myriad unintended consequences. Dams block sediment from supplying land downstream, contributing, for instance, to Louisiana's southern coast disintegrating. Dams also eliminate the shallow rocky habitat salmon need to spawn and block fish from moving up and downstream. Although dams are often built to supply water, they end up creating water haves and have-nots by moving water from one area to another. And they increase water demand among the haves by holding that "new water" in a big lake, giving people a false sense of bounty that encourages waste. Similarly, levees and seawalls protect one community but push higher water onto communities downstream or down coast. Levees also increase the scale of flood risk for the "protected" community by encouraging people to move into harm's way and by narrowing the floodplain, raising water levels. And by cutting off slow water from the land, they reduce storage underground, contributing to water scarcity.

By constraining natural water systems, we have claimed space to build towns, cities, and agricultural empires. But the idea that we can control water has always been a fantasy. Now that reality is sharpening into focus as our gray infrastructure—not designed to cope with the water extremes we are seeing today—is increasingly failing. These failures are prompting the water detectives to think differently. They are realizing that the natural systems we are destroying could be our salvation.

Human Blinders

To identify which land is a priority to conserve or restore, water detectives are trying to figure out what water did historically. That's a mystery to most of us because the human lifespan is out of sync with the scale of the changes we are causing. Consider my childhood home in what is now called Silicon Valley, once known as the Valley of Heart's Delight, a moniker made possible by water. In the mid-twentieth century, each spring heralded a flying carpet of white and pink blossoms atop acres of stone fruit trees: Bing cherries, Blenheim apricots, Burbank plums. By the time I was a kid in the 1970s and '80s, the valley was leaving behind the last vestiges of its agricultural past. Just a few small orchards remained, including my grandparents' hobby orchard that kept us in applesauce and dried apricots year-round. As I got older, the last of these orchards fell to strip malls, big-box stores, and retirement homes, and I was haunted by an aching sense of bereavement for a bucolic past I had barely known.

Yet despite my grief, I was incapable of fully understanding the scope of loss. The orchards I mourned had replaced an earlier, wilder landscape of native plants and animals. Oak savannas and woodlands, populated by tule elk and wolves, were shaped by the Ohlone, peoples who lived—and still live—in the South Bay. Their prescribed burns and coppicing cultivated the acorns and camas they liked to eat. Spanish explorers in the eighteenth century dubbed the area *Llano de los Robles,* or Plain of the Oaks, and into the late nineteenth century, people still marveled at the size and numbers of these captivating trees. But to make way for the orchards on prime valley land, early farmers felled up to 99 percent of them.

My skewed sense of the landscape's "natural" state is an example of a phenomenon called shifting baselines, coined by marine biologist Daniel Pauly in 1995. Basically, our sense of place, our love for a landscape, our notion of how it "should" be, is tied to our earliest memories. Humans can't truly grasp how much we have degraded the natural world because our baseline—our concept of what's natural—shifts with every generation. For centuries, humans have been diminishing the natural world: the decimation of central Europe's forests in the eighteenth century, the disappearance of grizzly bears from California in the early twentieth, melting glaciers in Switzerland, Iceland, and Peru in our current era. Today we live in a world that contains a fraction of the vast abundance of other species Earth once held—10 percent is the educated guess author J. B. MacKinnon offers in his 2014 book *The Once and Future World*.

Shifting baselines distort our understanding of water as well. The orchard landscape that so captured my imagination as a kid had already reshaped local hydrology in fundamental ways. To keep the valuable stone fruit crops viable during the area's dry summers, early farmers pumped groundwater enthusiastically: wells drilled increased seventeen-fold annually between 1892 and 1920. The hurrah was short-lived. Water levels plummeted and, without water pressure from beneath, the land surface sagged over many miles—a phenomenon that geologists call subsidence. Subsidence can break infrastructure such as buildings, pipes, roads, and irrigation channels. In downtown San Jose, the land dropped thirteen feet. This alarming development inspired early experiments across the area with groundwater recharge—purposefully moving water underground for storage.

A Time to Deconstruct

Today around the world, people are increasingly recognizing the trouble our water interventions have caused, and the call for nature-based solutions is growing louder. In 2018, the United Nations published a strategic report called *Nature-Based Solutions for Water*. Even the US Army Corps of Engineers, infamous in environmental circles for its muscular engineering of rivers and wetlands, now has an "Engineering with Nature" initiative that blends in some greener solutions. The World Bank is also advocating for nature-based solutions to water management, as are those historically famous water engineers, the Dutch. (In fact, the Dutch have created an Agenda of International Water Affairs to share their hard-won water wisdom through partnerships with other governments around the world— and to potentially win contracts for Dutch governments, research institutes, NGOs, and companies. I came across their work in multiple places.) But because we're already so invested in gray infrastructure and built-up cities, these institutions are not calling for Slow Water to replace our engineered water systems; rather, they would augment them, creating a blend of engineered and natural solutions.

However, we likely need a more radical revision, given the velocity at which climate change and its accompanying water disasters are accelerating. The upside: with problems comes opportunity to significantly change how we do things. Existing water infrastructure is aging out; many cities lose 40 percent or more of water to leaks. In 2021, the American Society of Civil Engineers gave the ninety-one thousand dams in the United States a nearly failing "D" grade. In spring 2020, a dam in Michigan failed and another was compromised. It's an international problem. Most large dams around the world were built between 1930 and 1970, with a design life of fifty to a hundred years. A dam in Kenya burst in 2018, killing nearly fifty people. Also, the water infrastructure we've built wasn't designed to accommodate the dramatic swings we're now seeing in precipitation and flows. China's massive Three Gorges Dam almost overflowed during 2020's summer rains, threatening tens of millions of people who live downstream.

Of course, we wouldn't want to lose the benefits of clean drinking water from the tap and sewage treatment. We will keep those systems. But increasingly, cities and towns are also uncovering stretches of buried creek (called daylighting) or funneling storm runoff into bioswales, those ditches planted with water-tolerant plants. For these efforts to work well, though, they can't be little add-ons here and there.

That's because, when we limit space for water in one spot, it pushes into another place within the watershed, the land area in which water drains to a common outlet like a river or ocean. Reclaiming one small area of floodplain won't decrease flooding much if the rest of the river remains leveed. Such small-scale thinking makes it hard to imagine how Slow Water projects can cope with the regular deluges we're now seeing. That's why, in a fundamental shift from the centralized gray-infrastructure water projects we've been building, water detectives aim to make space for Slow Water across the whole landscape. The vision is to deploy a series of small projects stitched together, so water moving through them can better function as the living entity it is. More is more.

Although an ever-growing number of successful Slow Water projects are popping up around the world, planners and funders, advised by engineers, still largely default to the gray infrastructure water projects they know—despite the damage they cause and their tendency to exacerbate the problems they are meant to solve. Witness the dam-building boom currently underway in Asia and South America.

Yet people are starting to see the limitations of those big engineered projects. Both Boston and New York City have considered and rejected giant flood barriers—mostly because they likely won't work. And a growing body of scientific research is showing decision makers that nature-based solutions can solve the problems at hand, often with extra benefits and cheaper price tags than gray-infrastructure approaches.

*

A big part of shifting to water management that collaborates with nature is giving up the illusion of control. Green infrastructure is not static or predictable like concrete: nature is messy. Water rises and falls. Plants sprout, live, and die. Mud is exposed. Although these spaces can be beautiful—perhaps more beautiful than, say, a dam—people might not always like what they see. We will need to learn to accept a dynamic environment, seeing riverbanks and coast-lines as flexible living spaces that are periodically inundated, filled with plants, or muddy. Widespread acceptance of Slow Water will require an attitude adjustment in the way we think about this vital compound, from commodity or industrial input to partner, friend, relative, life.

Landscape designer Yu Kongjian is an international leader in the Slow Water movement and an early proponent of China's Sponge Cities initiative that seeks to make urban zones better able to ab-sorb rainfall. He calls this shift in perspective "big-feet aesthetics," a counter-reference to when Chinese considered bound, tiny feet on women to be beautiful. Feet just a few inches long were desirable because they effectively hobbled the women, a sign that they were too rich to work. Similarly, coiffed lawns and heavily cultivated or-namental plants require a lot of added water and industrial inputs and often don't do much for the local ecosystem. "Now we need to find big feet attractive," he tells me, meaning embracing plants and landscapes that do important work.

Accepting big feet might be easier than we think. Western culture has a history of trying to elevate humans above nature, yet we are an-imals. That's why separation from nature can make us depressed and anxious and why many people feel grief as wild species and natural spaces disappear. "Trees, fungi, salamanders . . . these are our blood kin, if you believe Darwin," says David George Haskell, a biologist at University of the South and author of beautiful books about eco-systems and nonhuman life. Slow Water projects such as a reclaimed floodplain in a city center make space for nature where we live.

A societal shift toward accepting ongoing change in our local en-vironment is also a partial return to when humans lived closer to the land and knew intimately every shift in nature's mood. That kind of attention may seem quixotic to those of us caught up in modern life. But there are degrees of participation, and engaging is not generally a burden. Such attention can become a meditation on the inevitabil-ity and beauty of change. In making space for our landscapes to be more flexible, in reacquainting ourselves with how natural systems work and how critical they are to every human endeavor, we lay the groundwork for the most important adaptation: enhancing our ability to go with the flow.

Given that each place is different, each project requires a bespoke approach that takes into account its unique ecosystem, hydrology, geology, topography, soils, weather, human needs, politics, and culture. As Yu puts it: "Every patient needs a different solution." For that reason, the Slow Water strategies people are deploying in this book are intended to inspire rather than explicitly prescribe. Each of them questions the ingrained assumptions reflected in the highly engineered way we manage water and shows that another path is possible.

In my place, California, the terrifying drought had one upside: it galvanized people to make radical changes in how water is governed, allocated, and stored for future droughts. Now water detectives are trying to better understand water's relationship with the subsurface, a tableau created in geologic time, as a first step to storing more water underground. UC Davis hydrogeologist Graham Fogg thinks a secret weapon is a special geological feature that could serve as a super pipeline to aquifers—"if you knew about them," he says. The trouble is that only three have been discovered. But scientists believe that every tributary coming down off the Sierra Nevada mountain range has one. Now Fogg and colleagues are on the hunt.

Water in Geologic Time

HOW ANCIENT RIVERS CAN
HELP EASE DROUGHTS

Tens of thousands of years ago, California's Sierra Nevada wore on their shoulders the ancestors of today's rivers. The waterways flowed down from the highlands to meander across the plateau of the state's Central Valley, a nearly flat expanse about four hundred miles long and seventy-five miles wide cupped between the Sierras to the east and the Coast Ranges to the west. Upon this vast plain, the rivers formed languorous braided ribbons of milky blue. The strange hue came from silt and clay ground by ice into a fine powder called glacial flour that floated suspended in the water, the particles absorbing purple and blue light from the sun. You can still see this phenomenon today in mountainous places where rivers are fed by glaciers, such as the Kicking Horse River in British Columbia, Canada, or the Lhasa River in Tibet. But not in California.

The glaciers then cresting mountains around the globe held captive massive amounts of Earth's water, so sea levels were nearly four hundred feet lower than today. As the ice melted, runoff scoured the land and rocks, pushing downstream the loosened heavy gravel, sand, and silt, called glacial outwash. Due to climate, glacial processes, and the lower sea level, the rivers cut valleys a hundred feet deep and a mile wide through sediment on the Central Valley floor. Later in the glacial cycle, as sea levels rose and streams flattened out across the valley, they lost momentum and slowed down. Lacking enough energy to continue traveling downstream, the gravel and sediment dropped out of the water column, backfilling the cut can-yons with this coarse material.

Over the last million or so years, glaciers crept down from the north, settled atop the Sierra Nevada, then retreated multiple times. Four of these events left major marks on the Central Valley. The rivers that glaciers spawned deposited sediments, cut through sed-iments, deposited gravel, and laid down more sediments, leaving imprints of this icy ebb and flow below the surface of the land we see today. In some cases, the gravel-filled valleys stacked one atop another. In others, they landed askew, as rivers sometimes jump to another course, a phenomenon called avulsion.

Water's ancient negotiations with rock and soil created these his-toric riverbeds, called paleo valleys, that lie hidden underneath us. Because they are more permeable than surrounding material, they are still the pathways that water wants to travel underground. Only three have been discovered so far in California. Hydrogeologist Graham Fogg has a dream: to find more of these paleo valleys and use them as giant drains that can absorb water from today's heavier winter storms and store it underground. A better understanding of water's relationship with buried geology would allow us to use those natural systems to both protect homes and businesses from flooding and sock away big reserves to help us survive today's longer droughts.

Fogg has become one of my go-to sources on groundwater in re-cent years, as other water experts routinely point me in his direction. The subject is both an expertise and a passion, reflected in his Tesla's vanity plate that reads "GRD H$_2$O." Officially retired but still hard at work advising grad students and reviewing new papers, Fogg can't let go of the mystery beneath our feet. "My wife thinks I'm incapable of retiring," he tells me on a mini field trip near Sacramento to see subtle surface clues that point to underground features.

This water detective has spent his career thinking about water's se-cret life in the underground. In a landmark 1986 paper, Fogg argued that what lies below us is more varied than people assumed—clay here, sand there—and that connectivity between coarser sections creates paths that water prefers. Although porous pathways make up only a small percentage of the subsurface worldwide—about 13 to 20 percent—surprisingly, they still interconnect extensively in 3-D. For this reason, paleo valleys, as shafts of major flow, are the most significant features underlying California's Central Valley, Fogg explains to me. While most deposits of underground water occupy former stream channels, the paleo valleys born in the most recent glaciations have superpowers for recharge: they are large, their unusually coarse gravel makes them highly permeable, and they are relatively shallow. To indicate how they were formed, cut, and filled, scientists often call them incised-valley fill.

To improve water security during this age of climate change, storing water underground "is a no brainer," Fogg says. Yet despite his success in getting people to think about the complexity of what's underground, putting that knowledge to practical use—by finding and using paleo valleys to rapidly funnel water underground—has been a hard sell. For decades, few people were listening. But recently, that's beginning to change.

Water Security in the Underground

California has historically cycled between droughts and floods, but today, climate change is making both phases more intense. The 2011–16 drought and the atmospheric river storms of winter 2016–17 are harbingers of patterns predicted to intensify. As of this writing in mid-2021, another drought was already underway. Human population in the state has doubled since 1970, to nearly forty million today, raising the stakes for these wilder swings. More people are living in the path of floods and are in need of water during droughts. As the state's previous drought wore on and meteorologists began to predict winter 2016–17's deluge, water managers, traumatized by the years of want, saw opportunity. Their thinking began to align with Fogg's as they wondered: Can we capture the extra water that's coming and save it for the next dry spell?

The question is all the more urgent because a key source of water storage that California has long relied on—the Sierra Nevada snow pack—is disappearing. During the twentieth century, Sierra snow supplied about 30 percent of the water consumed each year by California's residents. Snow storage is conveniently timed, as it melts slowly through the spring and summer, just when it is needed most. But now warmer temperatures are bringing more rain and less snow, on average. Climate scientists predict that by 2100, the snowpack could shrink by four-fifths or more. As precipitation falls increasingly as rain instead of snow, there will more flooding in winter and less water supply in summer.

This is not just a California problem. Loss of snowpack and dwindling glaciers threaten the water supply of at least two billion people around the world: Asians who rely on the Himalayas, Europeans dependent on the Alps, South Americans served by the Andes. The endurance of those societies will depend on new ways to capture heavier flows, both to protect homes and businesses and to store water for later use.

In California as in other places, new reservoirs cannot solve the problem. Remember, we've already dammed nearly two-thirds of the world's mightiest rivers. While people have long stored water underground, California water managers are now planning to do so on an unprecedented scale. It's a solution that could be used in seasonally dry places around the world. Paleo valleys aren't required. Early projects have temporarily flooded farm fields to move water underground, or reclaimed parts of floodplains. But land in winter can become saturated. Using paleo valleys, because they allow water to move underground so quickly, could be the most potent method to capture the big winter storms to come.

Paleo valleys are a special type of the underground water features called aquifers (Latin *aqui*, "water," and *fer*, "bearing"). Aquifers are not big pools of water, but deposits of gravel, sand, and other porous material that water can move through more easily than surrounding rock or clay. Clay, which water can move through, but much more slowly, is sometimes called an aquitard (*tardus*, "late, slow"). Aquifers are cradled and contained by impermeable layers, such as rock. Some also lie beneath rock and are not connected to surface water.

Broadly speaking, modern people's relationship with groundwater has been mostly take, little give. In California, an average year saw people pumping groundwater from aquifers to meet nearly 40 percent of demand. When state and federal agencies reduced water deliveries during the last drought, farmers pumped at a furious rate, and that figure hit 60 percent. There were cases of farmers spending up to $400,000 digging wells to reach water tables that had plummeted two thousand feet below the surface. Some cities and farmers are 100 percent dependent on groundwater. But nature is unable to refill that supply at the pace people are extracting it, so water levels underground are falling.

This is an international problem too. Around the world, wherever good soil, sunny weather, and capacious aquifers align—the central United States, South Asia's Indo-Gangetic basin, the North China plain—people are pumping out groundwater faster than rains can soak into the soil to replenish it. That's the case for more than half of Earth's largest aquifers, according to a 2015 study using NASA satellite data.

The majority of the fresh water that humans tap worldwide goes to irrigate crops, and most pumped groundwater helps farmers fill the world's food baskets. If we run out of groundwater, a worldwide famine could result, given that the UN Food and Agriculture Organization estimates we need to *increase* global food production by up to 70 percent to feed the projected 9.7 billion people in 2050.

In California during the long drought, the falling water table was a sign of a situation so desperate that it spurred the state to regulate groundwater for the first time. Previously, California regulated surface water but treated groundwater as a property right: you could dig a well on your land and pump with abandon. If that drained your neighbors' wells, their only recourse was to take you to court. Treating surface water and groundwater as two separate supplies was a willful misunderstanding with a convenient implication: it allowed people to think of groundwater as a reserve bank of extra water to dip into when rivers ran scarce.

But in fact, groundwater isn't extra. Lakes, rivers, and streams on the surface and aquifers below are parts of the same water system. They are intricately linked by gravity and hydraulic pressure. A full aquifer can help feed a river's flow, pushing water up through its

streambed during the dry season. And vice versa: when groundwater levels fall, river water can filter down to replenish it, leaving less
water on the surface for people and endangered fish.

Pumping groundwater to replace surface water is therefore a
sleight of hand: it can deplete surface water, contributing to the very
problem that spurs farmers and cities to pump in the first place. Bad
groundwater management can destroy a hydrological system. For
example, in Southern California, water tables have fallen so far belowground in some places that they've become functionally discon-
nected from the streams with which they once exchanged water.

Yet perhaps that abuse has a silver lining: now there's a lot of
space to store water underground. (While overpumping can sometimes cause subsidence, as early orchardists saw in San Jose, leading
to permanent loss of capacity in an aquifer, that is not a universal
phenomenon.) In the Central Valley, after decades of overpumping, depleted aquifers have unused capacity three times that of the
state's 1,400 reservoirs. And storing water underground is a bargain
at roughly one-fifth the cost of building reservoirs.

But first we need to learn more about what's going on down
there. California's long political aversion to managing groundwater
is partly why the anatomy of the subsurface remains largely unex-
plored, says Fogg, and that ignorance leaves water managers flying
blind in establishing new ways to handle water. Studies in the works
today will show us the best places to spread water on the surface
to move it underground quickly, help us understand where we can
expect to find that water later, and learn how to avoid polluting it.

"The Least Wild Landscape Imaginable"

Driving the I-5 down the middle of the Sacramento Valley, my eyes
track the relentless geometry of row crops and orchards: sunflowers,
tomatoes, peaches, walnuts, olives. A summer pit stop in Willows or
Vacaville feels like stepping into a furnace. It's as far from glaciers as
one can imagine, making it difficult to conceive of their signature
underfoot. The soil is dry too, save where sprinklers dispense their
life-giving water, because California typically receives almost no

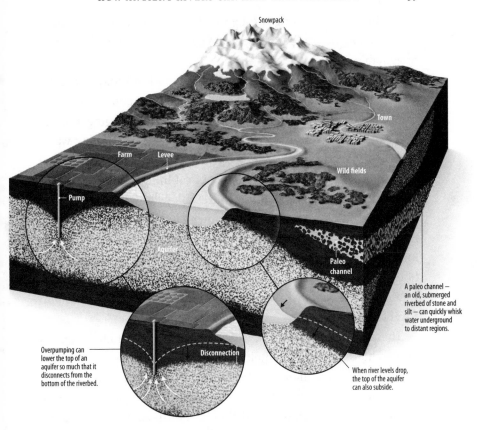

Snowpack

Town

Farm   Levee

Wild fields

Pump

Aquifer

Paleo channel

A paleo channel — an old, submerged riverbed of stone and silt — can quickly whisk water underground to distant regions.

Overpumping can lower the top of an aquifer so much that it disconnects from the bottom of the riverbed.

Disconnection

When river levels drop, the top of the aquifer can also subside.

FIGURE 2.1 Illustration of groundwater systems. Illustration by Emily Cooper

rain from April to October. The Sacramento Valley is the northern section of the Central Valley, named after its biggest river. In the same vein, the southern section is known as the San Joaquin Valley.

Eighty percent of water used by humans in the state goes to irrigate California's $38 billion crop industry, which supplies more than a third of US vegetables, two-thirds of its fruits and nuts, and international markets. This bounty is made possible by tapping groundwater and micromanaging a vast complex of water-engineering projects: dams, reservoirs, aqueducts, canals, levees, and pumps that have fundamentally changed the natural hydrology of the entire state and caused countless unintended consequences. The San Joaquin River is so diverted that, for more than a half century, it often ran dry for sixty miles. Beyond that, starting at the Mendota

Pool, the water coursing through its banks was not even its own. One water expert called the Central Valley "the least wild landscape imaginable."

It takes a strong imagination for a valley visitor today to picture what once was. Likewise, it would be hard for someone here before the Gold Rush in the mid-nineteenth century to imagine a future with depleted groundwater. Back then, the Central Valley was a vast, seasonal floodplain created by the mighty Sacramento and San Joaquin Rivers and their tributaries. Local Indigenous people told early European arrivals that the entire valley flooded almost annually from winter rains and spring snowmelts. One settler saw flood marks on trees ten feet high—proof that the flooding created "one immense sea," recounts historian Robert Kelley in his fantastic 1998 book *Battling the Inland Sea*. The water then drained slowly back into waterways, ultimately making its way to the delta that empties into San Francisco Bay.

These regular infusions of water spawned a riot of life. Grizzlies hunted herds of tule elk, antelope, and deer, as the ungulates grazed on grasslands of wild oats, dotted with charismatic valley oaks. The rivers and streams jumped with salmon and other fish. Vernal pools, home to fairy shrimp, reflected clouds of birds so dense they darkened the sky. Along the rivers, thick bands of forest held sycamore, cottonwood, willow and ash. Where standing water kept trees from growing, miles-wide marshes of tule reeds stood ten to fifteen feet tall, "impassable for six months out the year," writes Kelley.

Tragically, many Indigenous people in the valley—home to various tribes, including Miwok and Yokut—died from diseases brought by Europeans, leaving the flatlands open to new settlement. The newcomers quickly transformed the landscape, hunting the elk and antelope nearly to extinction, replacing them with sheep and cattle that may have hastened the demise of native plants by spreading seed from nonnative grasses. Settlers felled the valley oaks to clear the land for farming and built "permanent" homes and businesses right on the riverbanks.

Almost immediately, the winter after the Gold Rush began in 1848, the Sacramento River showed its true nature, inundating the new homes and farms with its annual floodwaters. According to *Battling the Inland Sea*, "Houses were toppled, businessmen watched thousands of dollars in inventory wash out of their doors." Beginning a few years later, hydraulic mining for gold in the foothills exacerbated the destruction. Men trained high-powered hoses on riverbanks in the foothills to loosen the soil in search of the precious metal, destabilizing mountainsides that slid down into the valley when winter storms came. In January 1862, the valley was once again a massive lake, and the new governor, Leland Stanford, who founded the university that bears his name, had to travel by rowboat to his inauguration and return to the Governor's Mansion in Sacramento through a second-story window.

Yet neither the settlers nor their government took repeated flooding as a sign "to get out of nature's way" and retreat to higher land. Instead, as Kelley puts it, "their instinctively activist impulse was to . . . force nature to behave as they wished it to."

The Sacramento River resisted control. It took seventy-five years of trial—and a lot of error—for people to successfully confine most of its floods to a narrow channel. Part of the difficulty was that the Sacramento rose much more quickly than the Mississippi, a river with which American engineers had longer experience. The Sacramento is a big river. Although late-summer flows averaged about three thousand cubic feet per second, in flood it could swell to as much as six hundred thousand cubic feet per second, Kelley reports. "The river's channel could never contain within its natural banks the huge flows of water that almost annually poured out of the canyons of the northern Sierra Nevada."

This epic fight with the great rivers of the Central Valley laid the groundwork for a radical replumbing of the entire state—arguably the apex of water-engineering hubris. That quest for control and water redistribution, primarily from north to south, created the California we know today. To a large extent it has encouraged people to live and grow food in places without enough local water to support the numbers and output currently there. This wasn't news: author

Mary Austin wrote a book in 1910 about Southern California called *The Land of Little Rain*. That precarious balance raises the question whether the state should be exporting crops to more water-rich areas.

Yet the domineering infrastructure that has worked, more or less, for a century may not hold much longer with climate change, population growth, and excessive groundwater pumping. The engineering has deprived the valley of its slow water phases, reducing nature's ability to refill aquifers. And it has driven many of the valley's native plants and animals to near extinction by drying up vernal pools and wetlands and dropping the groundwater table. As the state begins to expand groundwater recharge, it could help these struggling species by allowing water, once again, to stall on the valley floor.

Percolation Station

Actively moving water underground is not new. It's an ancient human practice, still done in arid rural areas—including in India, Africa, the Middle East—where farmers contour their farmland in shallow basins or trenches to allow rainfall to soak into the soil. The corralled water doesn't even have to filter down to an aquifer to make this strategy useful. By lingering in the soil, rainfall can help crops grow with reduced need for irrigation.

Today's water managers have increased the capacity of this basic idea with percolation ponds and injection wells. They call this managed aquifer recharge (MAR). Within California, there are early examples of recharging groundwater. The need to do so was driven by heavy pumping that caused subsidence, seawater intruding into fresh groundwater, or one landowner suing another when wells ran dry.

My childhood home of Santa Clara Valley was a leader in recharge, motivated to counteract the dropping water tables and sinking land caused by its fruit-production boom. Local leaders created recharge basins—including some right along creeks—where water could slow down and soak into the ground. Just as beavers dam a stream to create a pool, humans built partial walls across creeks in the west Santa Clara Valley in the 1920s, using burlap sacks filled with dirt, according to the California History Center. The structures slowed the creek's flow, creating ponds alongside that gave the water time to percolate underground. Earthen dams followed, such as the thirty-four-foot-tall Vasona Dam across Los Gatos Creek that sits at a topographical low spot in a beloved town park. When heavy rains come, the park floods, allowing extra recharge without threatening surrounding homes or businesses.

Since those early efforts, Valley Water, the local water district, reports that it has built recharge basins alongside more than ninety miles of local creeks. It has also installed three hundred acres of freestanding percolation ponds, dug-out depressions built above porous ground. It fills these with water piped in from nearby creeks when they run high or from state and federal water projects that deliver water from further north. For smaller recharge sites in dense urban areas, the district can inject water underground via wells. Recharge really shows its value during droughts. The supply is a buffer when imported deliveries are curtailed and gives the district time to implement community conservation programs.

Today there are at least 1,200 MAR projects in sixty-two countries, according to the International Groundwater Resources Assessment Centre based in Delft, the Netherlands. Several cities and towns in Arizona and Southern California have made recharge a significant part of their water management.

*

Like many other Californians, I had not given a lot of thought to what's lies under the ground beneath my feet. If asked, I'd have come up with soil, rock, sand, clay, possibly some pockets of oil, gas, minerals, water. That's not wrong, but it leaves a lot out. I learn more when I meet up with Graham Fogg northwest of Sacramento for that mini field trip on a hot May day in 2021. A mild-mannered man with white hair and glasses who clearly loves teaching, Fogg pulls up a satellite view map on his Tesla's dashboard tablet to show me the visible imprint of past creeks and other water features.

"The subsurface has its own anatomy and physiology," he explains, and its shape affects the way water moves underground. Although the Central Valley is one of the biggest aquifer systems in the world, in fact, 65 to 80 percent of it is silt and clay and not very conducive to moving water. That's why finding the paleo valleys that are close to the surface is so important. The ancient waterways now filled with porous gravel and sand can still provide the main path for water, as they did during their time on the surface.

Throughout the clay-dominant substrate, imagine that the paleo valleys are like mainline blood vessels of the groundwater system, with myriad branches tentacling off like capillaries. As the groundwater moves through these channels filled with sand and gravel, it is also moving in and out of the surrounding clays and silts. Fogg extends the metaphor of our body tissues: "The fluids in your body move through veins and arteries relatively quickly. But most of your body is these soft tissues that are primarily water; the water moves in and out of these more slowly, by molecular diffusion and other processes."

In fact, a complex water cycle is happening underground. In school we learn the mind-blowing fact that just 3 percent of the water on Earth is fresh water, and only about 31 percent of that is liquid, not ice. Amazingly, 96 percent of that liquid fresh water is underground. It resides there partly because gravity pulls it down. But it's more complicated than that, Fogg says.

Groundwater circulates deep into Earth, recharging and discharging on a local scale but also regionally. "That's how you get fresh water down thousands of feet in a place like the Central Valley," Fogg explains. Before people began aggressively pumping groundwater in the mid-twentieth century, it discharged naturally to streams and wetlands. That process is now greatly diminished, and in some places eliminated. Today, most groundwater that would otherwise recharge the system is pulled back out by people, from wells.

It's a bit confounding to think about how gravity, which pulls things toward the center of Earth, can move water up to the surface. Hydraulic pressure is also involved. Fogg suggests that I think of a U-shaped tube with a stopper on one side, filled with water. Due to the force of the water pushing upward on the stopper, the water level is higher on the other side. When water enters an aquifer from a higher elevation, he explains, it puts pressure on the subsurface, which can drive the water deep down. But when that water reaches certain depths, the pressure can drive it vertically upward. This phenomenon feeds springs, perennially flowing creeks, and most wetlands, which would not survive without upward-flowing groundwater to sustain them. "These are big systems that have a lot of things going on," he tells me, with a tone of fond awe.

Just as Joel Pomerantz in San Francisco spots clues to underground water in the urban landscape, Fogg has a good idea of what lies beneath us by observing the lay of the land. Topography—the hills and valleys—is a loose indicator of the path water takes underground (although natural patterns can be confounded by human engineering). On our field trip, he shows me signs of a large paleo valley, created by an earlier avulsion of the American River. Like many ghost imprints of water on the modern landscape, the things he's pointing out are subtle. Without Fogg's guidance, I would look past these features without realizing their significance. Piles of cobble—stones roughly the size of my hand, now partly grown over with grasses—are dumped refuse from hydraulic gold mining. A small dry creek through a field is lined with similarly sized rocks. Fogg tells me that paleo valleys also have this architecture because streams and rivers can only move such large rocks during really big flows. A gentle ridge of land in the distance could be the edge of an old river channel, because in buried valleys filled with sediment and gravel, the coarser material is resistant to erosion.

Fogg's body of work brings together two related fields, hydrology and geology; he has degrees in both. While hydrogeology is supposed to be the study of water underground, Fogg was troubled that practitioners strongest in hydrology, lacking data and understanding about the subsurface, assumed it was so chaotic as to be unpredictable and tended to ignore or misconstrue what happens there.

Early in his career, Fogg spent eleven years working with geologists at the Bureau of Economic Geology in Texas. To these specialists, the various textures and grain sizes of soils, the traces of ancient plants striped through the underground, are a language that tells the story of the events that laid them down. The reverse is also true: geologic history dictates where layers of finer sediment and gravel can be found.

Fogg understood that was the missing link for hydrologists. "What I saw in geology was, there's predictable organization to these systems, right?" The predictability inherent in that historical finger-print is critical to understanding what we can't easily see. "That's how I can say that every major stream coming off the Sierra Nevadas has a shallow paleo valley produced by the last glacial episode," Fogg explains. He would also expect to find similar features elsewhere in the world where mountains meet sedimentary basins and there have been glaciers during the last million years, such as the Ganges plain fronting the Himalayas. Fogg is counting on that predictability to help scientists find shallow paleo valleys more quickly and cheap-ly—a skill that could prove valuable as California water managers move to expand groundwater storage. Fogg calls them "Yosemite Valleys" for groundwater recharge because their qualities are as unique for recharge as Yosemite Valley is for its aesthetic beauty. That image has inspired water experts to suggest setting aside land atop paleo valleys as recharge preserves because of their public benefit.

Farming Water

When state and local water managers started planning large-scale recharge during the California drought of the 2010s, they lacked a clear picture of what lies underground. They thought maybe they could use some of the original valley floodplain, allowing water to linger on the vast acres of farmland that have replaced it. Flooding some of these fields in winter would be a loose approximation of those historic inland floods.

But understandably, farmers wanted to make sure that flooding wouldn't hurt their crops. And water managers wanted to make sure that fertilizers and pesticides applied to the crops wouldn't pollute the groundwater. Scientists conducted test floods that measured plant and root health, water infiltration rates, and pollution levels.

The results have been generally positive, recharging a lot of water without harming crops.

One grower was way ahead of the curve with this line of thinking. Don Cameron manages Terranova Ranch, a seven-thousand-acre farm in the San Joaquin Valley southwest of Fresno that grows twenty-five different conventional and organic crops, almost entirely with groundwater. With water tables dropping, he knew he needed to recharge water to continue farming there. The local Kings River Water Association allowed him to test recharge by taking excess water in winter 2011 and again in 2017 that its members weren't using. Cameron funneled the high flows via canals to fields both fallow and full of wine grapes, olives, almonds, and pistachios. It worked: crops were unharmed, and sensors showed that at least 70 percent of the water passed below plant root zones. In 2020 he expanded canals and pumps to be able to flood nearly his entire acreage, and his local water district won a grant to expand recharge on an additional six thousand acres of neighboring land.

Still, if recharging water on farm fields becomes widespread, keeping nutrients and pesticides out of recharged groundwater will require timing fertilizer applications away from flooding. These crop supplements have already polluted groundwater in parts of the state, especially in places with flood irrigation. Once groundwater is polluted, it's expensive and difficult to clean it.

Also, farm fields aren't always ready to recharge water when the winter rains come. Sometimes a crop is still maturing, or the land is an orchard, filled with trees who don't like their feet wet. Plus, capturing the winter and spring water available for recharge can be difficult during wet times when the land is already saturated and every reservoir is full.

These challenges are why Fogg is so excited about the promise of paleo valleys, which would sidestep potential conflicts with agriculture. Officials could create those proposed preserves, conserving the land above them for dedicated recharge. In the culture of California's heavily managed water system, it would likely mean building new canals or pipelines to deliver floodwaters to these optimal recharge areas. But first, scientists need to find more of them.

Hunt for the Wild Paleo Valleys

Paleo valleys were first described and identified by geologists starting in the 1980s, but it would be more than a decade before any were
found in California. That notable moment came in 1992, when Fogg's
doctoral student Gary Weissmann rushed into his office. Weissmann, who grew up in Colorado, attended UC Davis specifically to
work with Fogg, who was one of the few people at the time asking
questions about how underground formations shape the paths that
groundwater travels. They were kindred water detectives on a common quest. At their first meeting, Weissmann says, "we spent quite
a bit of time getting excited about the prospects of building a better
understanding of aquifers. We both geeked out in a big way."

Today Weissmann is a professor of hydrogeology at the University of New Mexico. But thirty years ago, he was studying an area
near the Kings River (incidentally, not far from Terranova Ranch),
trying to understand whether a pesticide applied to fields outside
of Fresno in the Central Valley might have traveled laterally underground to taint local wells. He was in the lab, going through the data
people collect when they drill wells to get water, oil, or gas. Some
wells are drilled with a core bit, a hollow tube that brings up a long
cylinder of soil. That soil core provides information about the materials they drill through: sand, clay, gravel. Each glacial cycle leaves a
recognizable soil in the subsurface—"the punctuation, so to speak,
of a glacial episode," as Fogg puts it.

While Weissmann worked, he recorded the soil strata in his
notebook, color coding the types of sediments along the edge of
the paper. He used red to mark ancient soils from the most recent
glacial cycle ten thousand to eighteen thousand years ago, known as
Modesto in California. Flipping through it, looking for a particular
well, he noticed his red marks were all lining up at similar depths.
He began plotting sediment types on a map, and the Modesto soils
ran in two parallel lines near the towns of Sanger and Del Rey. But
between them lay a gap—a different material.

"I thought, 'That's kind of weird,'" Weissmann recounts. He
looked at additional well logs from inside the gap and saw no red

marks there, indicating that something had cut it and moved it away. The valley was cut down into older sediment, called Riverbank, from the previous glacial cycle. That's why he saw no red-labeled soil in the channel cut between the towns.

"I also saw indications of pretty coarse-grained material like the rubble deposited by glaciers in paleo valleys," Weissmann remembers. "I got all excited." Running into Fogg's office, "I told him, 'I think we've got a valley here!'" Weissmann had discovered the first known paleo valley in California.

All these years later, Weissmann is still obsessed with the patterns that rivers lay down in sediment. "The deposits of rivers are beautiful," he rhapsodizes. "It's a pretty puzzle . . . to understand how all these pieces of the river got together." Among his favorite deposits are the sediments that water shapes into alluvial fans: think of the splayed shape of the Mississippi delta near the Louisiana coast, or the Mekong in southern Vietnam. Smaller streams also create fans, and ancient rivers long since buried have left this signature mark in the strata underground.

Fogg thinks the most useful paleo valleys for groundwater recharge are the most recent because they are closest to the surface— perhaps just a yard or two down. They would allow water to move rapidly underground compared with the vast majority of the surface area that has much slower infiltration rates due to the widespread presence of silt and clay. Cut through the sediment about sixteen thousand to eighteen thousand years ago at the glacial peak, these channels then filled with gravel as the glaciers melted from approximately ten thousand to fifteen thousand years ago.

To confirm that Weissmann's find was, in fact, a paleo valley, he and Fogg won a grant from the National Science Foundation to drill cores to see if they could find the cobble. Typically, as the drill bit spins, it penetrates sand, gravel, and mud pretty smoothly. "But when we hit those cobbles, the bit is bouncing up and down a few inches— which is significant—and you can hear it grinding around," Weissmann recounts. "That indicates to me that you've got at least fist-sized cobbles. It was pretty wild." With a regular bit, they were able to feel out about twenty-six feet thick of cobble. Paleo valley confirmed!

"The discovery . . . was just incredible," Weissmann gushes. "I was giddy about it."

Weissmann's 1999 publication on the find in the *Journal of Hydrology* was well received in the scientific community, as were subsequent papers. Fogg thought the discovery's usefulness for recharge was self-evident and would inspire a trend. "As an academic you think, 'OK . . . eventually more and more people will find these things and things will progress.' Well, it didn't. It basically sat there."

Three years later, Fogg went on a lecture tour of fifty-six campuses for the Geological Society of America. "I really pushed hard on the point that these features are major recharge resources and also major avenues for contamination." Still, the response was crickets.

The same willful blindness that kept state regulators from measuring how much groundwater was being pumped out because it was an inviolable property right made them incurious about how water moved underground or how groundwater and surface water interacted.

But Weissmann and Fogg continued to spread the gospel through their teaching. In 2005, one of Weissmann's grad students, Amy Lansdale Kephart, went through ten thousand well logs and discovered another paleo valley near Modesto. She built a 3-D map with the data and modeled how water would flow through it and how useful it could be for water storage. The Modesto paleo valley is about 0.4 to 1 mile wide and 10 to 98 feet thick. She found it could influence groundwater flow, attracting water into it or pushing water out of it, for around 12.4 miles on either side of the valley and for hundreds of feet in depth.

A student of Fogg's, Casey Meirovitz, found a third paleo valley near Sacramento in 2017. Then, in 2019, Fogg's student Steven Maples showed that it could accommodate almost sixty times more water than surrounding lands. "It's a shame that we still don't know where the rest of them are in California," says Fogg in his mild way. "This should be a priority."

Groundwater Regulation at Last

Finally, groundwater storage is becoming more important to state water managers. "It wasn't until the drought crisis hit that people began to look more carefully at this kind of thing and say, oh, yeah. Now maybe we see how important this could be," Fogg mulls.

The drought and the panic it induced also cracked open a door to finally get Californians past their political unwillingness to regulate groundwater. After decades of trying, in 2014, the state legislature passed California's most significant water reform in a century: the Sustainable Groundwater Management Act (SGMA, pronounced SIG-ma). The law finally acknowledged the reality of water physics: that groundwater and surface water are part of one system and need to be managed together.

SGMA requires that people sustainably manage groundwater basins—three-dimensional areas that include a watershed and the aquifer below. Neighbors in each basin must form a groundwater sustainability agency (GSA) and craft a plan—by 2020 for critically overdrawn basins and by 2022 for the rest—to achieve sustainable management by 2040. Because people have become accustomed to pumping groundwater, restricting extraction is unpopular. For that reason, the law encourages recharge because if people replenish groundwater, they can continue pumping when they need it.

That's true for the McMullin GSA, which includes Don Cameron's Terranova Ranch. Cameron, sometimes called "the grandfather of groundwater recharge," is busy convincing his neighbors that they can store more than two million acre-feet of water underground—twice the capacity of the local Pine Flat Dam. (An acre-foot is the quantity of water needed to flood an acre of land a foot deep. It pencils out to 325,851 gallons.) Storing water underground avoids the environmental impacts associated with building a dam and reservoir, he points out, and is much cheaper. In fact, groundwater recharge is now becoming so popular that people are fighting over floodwater and the State Water Board is mediating. "What was once a problem that no one wanted is now valuable," says Cameron.

Other places could use emptied aquifers in this way too. A recent study by researchers from the University of Texas at Austin found that depleted aquifers along the east coast of Texas have enough space to store two-thirds of the water from high-flow events from ten Texas rivers, reducing the impacts of both floods and droughts.

In California, there is enough unmanaged surface water statewide to resupply Central Valley aquifers, according to a 2017 study. The state's big engineered canals and aqueducts that send water from the north to fill irrigation systems in the south are underused in winter, when fewer growers need to irrigate. These channels could be used to transport excess winter water to San Joaquin Valley's empty aquifers.

But SGMA may also move people away from California's massive statewide water infrastructure toward smaller, local projects. Unlike surface storage, where people build giant reservoirs, Slow Water mimics the way nature stalls water on land across a watershed, so it requires thousands of grassroots projects suited to local conditions. A dam and reservoir are easier for people to grasp, Fogg observes: "The politician cuts the ribbon and everyone's happy." Diverting high-magnitude flows for recharge is more complicated. It requires cooperation among various people and organizations, getting access to the water, and moving it into position. But scientists are figuring out how to make this work. A 2021 study in *Science Advances* modeled floodwater quantities available for recharge in California under different climate scenarios and identified key infrastructure needed to store them. The authors found that under a low-emissions trajectory, water available for recharge would increase in 56 percent of the state's sub-basins during the period from 2070 to 2099, compared with 1976 to 2005. A high-emissions trajectory would see an increase in 80 percent of sub-basins. However, infrastructure limitations and policy hurdles must be overcome to make full use of it.

Creating multiple small local projects also tracks with the local management at the heart of SGMA. The only way the law could pass politically was to allow local basins to manage themselves. That's because farmers and their lobby are typically distrustful of Big State Government. (Although that's ironic, given that they benefit from state and federal largesse. As author Marc Reisner puts it in his classic tome on Western water, *Cadillac Desert*: "With huge dams built for him at public expense, and irrigation canals, and the water sold for a quarter of a cent per ton—a price which guaranteed that little of the public's investment would ever be paid back—the West's yeoman farmer became the embodiment of the welfare state, though he was the last to recognize it.")

Ultimately, SGMA pushes farmers and cities within a watershed to work together to meet sustainable usage levels—which could prompt Californians to live more within local water means.

Floodplain Power

To make the most of Slow Water solutions, paleo valleys should not be the only sites for recharge. Creating conservation areas along rivers' former floodplains is classic Slow Water and brings with it multiple benefits for various forms of life (more on that in the next chapter). The soil along rivers is typically fine grained, silt and clay, which doesn't allow for fast recharge. But if it's not built or planted on, it can remain inundated for longer periods of time; after all— and this cannot be overstated—*floodplains exist to absorb floods*. In a big enough area, "you can still accomplish a fair amount of recharge there," says Fogg, with added benefits of flood control and improved ecosystem health.

During the notably wet February 2017, I traveled to the Sacramento–San Joaquin Delta, where the mighty northern and southern rivers meet, meander, and ultimately spill into San Francisco Bay. The delta is the epicenter of California water, through which two-thirds of human-used water in the state flows. On a day in which sun is beating cloud, wildlife is enjoying the bounty of the recent rains. Several of the telephone poles I pass along the road are topped by Harlan's hawks, including one holding a wriggling mouse.

Life in the delta is tenuous for all parties. Here, land and water hold an uneasy détente. Over the past 170 years, people have remade this marshscape, pushing up earthen levees that grid the delta like fishnet stockings, cordoning off farmland from the rivers. Within lie reclaimed "islands" (actually sunken depressions, created by that peat oxidization). Aquatic wildlife here is not just deprived of water but also pulverized in the giant pumps that divert fresh water south toward Los Angeles.

I'm here to meet Josh Viers, a researcher at the Center for Watershed Sciences, and two people from the Nature Conservancy (TNC). The NGO led the levee setback project we are here to see in the Consumnes River Preserve. We pull off a two-lane highway surrounded by farmland at the eight-hundred-acre Oneto–Denier restoration site. Valley oaks, Fremont cottonwoods and wapato still hug the river. We walk atop an earthen levee, flanked on either side by floodwater three to six feet deep, soaking the feet of the riparian trees. Beyond them I see flooded fields that TNC is restoring to native habitat.

In 2014, TNC removed 750 feet of levee to allow the Cosumnes River to fill this part of its floodplain when waters run high. Because of the long drought, winter 2016–17 was the first real test of the levee removal. Graham Fogg was involved here too. He set up instruments to measure the groundwater recharge. When the floodwater receded, his students calculated that the flooding had recharged three times more groundwater than is typical from rain and irrigation. That winter's flooding raised local groundwater levels by seventy feet, says Fogg, to within two feet of the surface. The Oneto–Denier floodplain is relatively small; nevertheless, early estimates show that winter's storms resupplied roughly fourteen hundred to two thousand acre-feet of water, the latter enough for more than twenty-one thousand Californians for a year. Upstream, the work has inspired people to begin similar projects.

In fact, California's official flood-control policy morphed in 2017 from building levees to pushing them back, to reconnect major rivers with their floodplains and to give water room. The goal: to reduce flood risk while increasing groundwater recharge and wildlife habitat.

Because floodplains are meant to be periodically inundated with slow water, young fish, including salmon, evolved to spend time there en route to the ocean from their birth streams. The shallow waters grow algae, which feeds tiny crustaceans. For the salmon, "it's liquid protein" fish biologist Jacob Katz told reporter Robin Meadows for a *bioGraphic* story. By contrast, narrowed rivers, holstered behind levees, "are essentially food deserts," Katz said. Various studies of accidental and purposeful introductions of fish to floodplains have shown that those who chill on floodplains grow bigger— weighing as much as twelve times more!—than those stuck within channelized rivers.

From another spot nearby—Wimpy's Marina, a local institution outside the tiny delta hamlet of Walnut Grove—we hop into a seventeen-foot Gator Jon boat. Viers confidently navigates us down the Mokelumne River past a small, accidental levee break caused by the big rains. Rounding a bend, a curious scene appears. California sycamores atop the levee are interspersed with tractors and other farm equipment taking refuge above the water line. A great horned owl watches us from upper branches. We pass into a wide, shallow lake, where valley oaks seemingly grow out of the water near flooded farm sheds, old trucks, and the rusting curved tines of a hay rake.

In this temporary lake we come upon two of Viers's grad students collecting samples from a boat to study how native fish benefit from floodplain habitat. Like Katz's research, the students' findings could deflate the "fish versus farmers" narrative, in which farmers argue that environmental laws to leave enough water in rivers to sustain endangered fish deprive people of needed irrigation. That's because allowing floodwaters to reach out over their floodplains is better for wildlife than simply sending more water down the river, as environmental managers currently do when fish like the infamous delta smelt are in trouble, says Viers. When fish get fat during slow water phases on floodplains, they grow stronger and more resilient, with less need for dedicated water deliveries to keep them from going extinct. The same water that feeds the fish eventually eases back into the river to supply farmers, reducing the need to ration.

The state is beginning to recognize this value, passing a law in 2016 that declared watersheds to be integral components of California's water infrastructure, opening up new funding sources for projects like levee setbacks.

Taking to the Sky to Peer Underground

Still, paleo valleys hold the promise of capturing a far greater quantity of winter rains. And as drought perpetually threatens, water managers in the state are finally getting serious about understanding the underground. They need a way to step up the pace from months-long scouring of uneven data from well logs to more broadly mapping the subsurface to expose water's geologic fingerprints.

Enter the helicopter. California's Department of Water Resources is investing $12 million to map groundwater basins throughout the state using a technology called airborne electromagnetic surveys, or airborne EM. Rosemary Knight, a geophysicist at Stanford, led an early project to try to understand how groundwater moves, including through paleo valleys.

In December 2020, a pilot and a technician flew the first survey in search of paleo valleys over Tulare and Fresno Counties in the southeastern part of the Central Valley. Due to the pandemic, I couldn't go watch this notable flight. But Knight sent me video of an earlier flight using the same technique, conducted for other research. In it, the helicopter hovers while a technician attaches the instruments by ropes. The pilot eases the craft higher, and an awkward sight emerges: hanging from the helicopter is a box about the size of a portable generator; beneath that, a large hexagram of pipe dangles from several ropes, giving the impression of a giant lasso, frozen open. The lasso is a transmitter through which electrical current flows, setting up a primary magnetic field. It generates eddy currents at various depths beneath the ground, and that, in turn, creates a secondary magnetic field measured by a receiver.

The data is converted to color coding to show sediments with low electrical resistivity, such as clay, in shades of blue, and high-resistivity materials, including sand and gravel, in red. The image is 3-D, showing an area that in this case was about three hundred square miles with depths up to a thousand feet. Swirls and pockets of red amid the blues reveal where water flows most easily. Airborne EM has been used in other places around the world to peep into the subsurface. But until now, it has scarcely been applied in California.

Fogg thinks that airborne EM will be "game changing" for finding the rest of the paleo valleys in California. But after getting a general idea of where they are, people will still need to map detailed information about them either with higher-density airborne surveys or the old-school way, with data from drilled wells. The latter should be easier now too. That data used to be proprietary. Whenever the legislature tried to submit a bill to make the information public, it met an implacable wall of resistance from industry, Fogg tells me. Now, thanks to a recent law, it's available to everyone; but he's found it to be in disarray. "It needs a fair amount of analysis and triage" to be more useful for groundwater science and management. Fogg would like to see a state agency sort this data into categories of "good, bad, and ugly." In recent years, he has been advocating for the California Geological Survey to take up mapping for water resources, shifting from its historical focus on mining, oil, and gas to the preeminent resource today.

Pay Dirt

Water managers in California are now thinking more holistically about water, considering the complex relationships among surface water, groundwater, wildlife, and people. It's a shift away from the long-held "Mine! Mine!" mentality that has led people to wrangle over every molecule. Some earlier recharge projects are examples of this older outlook. One in San Joaquin Valley's Kern County treats aquifers as a water bank: if you make a deposit, you get an equal withdrawal. The problem is that in many geologies, water doesn't necessarily stay in a tidy underground basin, waiting to be pumped out when needed. Some of the water moved underground can be "lost" to the wider environment. Accepting some loss means moving from an ownership mentality to one focused on sustainability, in which recharging water supports ecosystem health, so fewer human interventions will be required to stave off collapse.

That mental shift may not be as hard to make when we recognize that gray infrastructure such as dams are not immune to water loss. Reservoirs behind dams lose a lot of water to evaporation. Just how much is difficult to measure, but globally the loss is estimated to be greater than the combined consumption from industrial and domestic water use. A thirty-year study found that the average water lost to evaporation each year from 721 reservoirs across the United States was nearly equal to the nation's annual public water supply. Managers also sometimes have to release water ahead of big storms to make room for high flows. "Loss" is already part of the status quo.

South of San Francisco, UC Santa Cruz hydrogeologist Andrew Fisher and his colleagues have pioneered a way to pay farmers for recharge services that recognizes that reality. Here along the coast, the rolling hills and farmland supply artichokes, berries and leafy greens worldwide. But there is little surface water; growers rely almost exclusively on pumped groundwater. In the 1980s, overpumping was already a big problem, allowing salt water to move in underground from the sea. The state then created the Pajaro Valley Water Management Agency and empowered managers to charge water users for the groundwater they pumped.

Fisher is working with the local resource conservation district on a project that diverts excess rainwater from fields and surrounding foothills into a four-acre dedicated recharge basin. In the rainy winter of 2016–17, his team recorded about 140 acre-feet of infiltration. Over nine years of tracking, the average they collected in nondrought years is about one hundred acre-feet. Since then, they have developed a larger project nearby and are in various stages of preparation for others. The first two projects are built over old stream channels, past tributaries of the Pajaro River. The sediments are coarser than the surrounding floodplain deposits: "Windows of infiltration and recharge, if we can find them," says Fisher. It's a similar principle to Fogg's paleo valleys, but the latter are bigger, even coarser, and older because they were formed by glaciation.

The stormwater infiltration earns farmers a rebate on groundwater pumping charges, much as utilities credit customers with rooftop solar panels for surplus power that they send to the grid, a policy called net metering. For the initial operation of this recharge net metering program, the Pajaro Valley Water Management Agency has been crediting farmers 50 percent of the infiltration rate against their future groundwater pumping costs. That conservative figure reflects the reality that some water would have seeped underground even without their efforts and that some of the infiltrated water will travel into the wider hydrological system before it can be pumped back out. But that loss is actually a "soft benefit," says Fisher, because it helps push salt water back toward the ocean, keeps the soil moist (reducing the need to irrigate), and maintains higher groundwater levels. "Instead of arguing about who owns water, why don't we claim the win based on the hydrological benefit?" he asks. "It's everybody's water because it goes into the basin that everyone is pumping from. Everybody wins."

A communal view toward water management is made more likely thanks to the flexibility that SGMA gives each groundwater agency to incentivize recharge in ways they feel are fair and that reflect cultural differences from basin to basin. Some of them may also decide to shift to less water-intensive crops or take some marginal farmland out of operation and return it to wildlife and groundwater recharge. When communities are empowered, people feel more invested and committed to the measures they agree to take.

They may even feel pride in stewardship. As the studies of fish on floodplains show, the same water can serve both wildlife and people—often better than when smaller portions of different water are allocated to each. Shifting from a scarcity mindset to one of shared abundance is hard; it goes against our dominant culture. But scientists who study microcritters are learning more about how a healthy ecosystem supports a healthy water supply (see chapter 3).

As for Graham Fogg, his decades-long dream of using paleo valleys for recharging water is finally inching closer to reality. But many land-use decisions are local, and much work remains to convince decision makers. On our field trip, Fogg shows me the Folsom South Canal, a straight concrete channel filled with water diverted from the American River. It goes largely unused, he tells me, as agriculture has moved on and houses haven't yet been built. Just beyond the canal lies an undulating field that rests atop the discovered Sacramento paleo valley. This proximity of underused water to paleo valley makes this area a prime spot for a conservation recharge zone—but it's private land, and the city wants to grant permits to developers to build houses.

Fogg may be a dreamer, but he's also a pragmatist. Folsom South Canal is owned by the federal government, which also owns a narrow strip of land alongside the canal. Injection wells drilled on the federal land could be a compact way to move canal water into the paleo valley. He has convinced a member of the Sacramento Area Flood Control Agency to work with the US Bureau of Reclamation to seize this opportunity. "There's plenty of space to store water. You could build the equivalent of another Folsom Lake," he says, citing the reservoir just up the road that holds nearly a million acre-feet.

Still, he's encountering resistance. Even water scientists and engineers—who are accustomed to drilling wells in places where they can tap groundwater—may have a psychological block to thinking seriously about exploiting these special paleo valleys for recharge. That's because, to most people, "the subsurface is still rather opaque and mysterious," Fogg muses. "Much of what goes on there is not very accessible to human consciousness." That's why a demonstration project could be revelatory, showing just how much more water people could recharge in the buried valleys compared with standard basins used for recharge.

Fogg remains hopeful, citing the enthusiasm of several people in the state's Department of Water Resources to identify these special features for recharge projects. Several years ago, in a speech to accept an award from the Geological Society of America, he revealed the quiet confidence that has kept him faithful to his vision during the decades in which few seemed to understand or care: "Decide in your head and gut what you think is the right path, and never give up—and of course, never stop listening."

· 3 ·

From Megadams
to Microbes

WATER'S RELATIONSHIP WITH
TINY FORMS OF LIFE

Dressed in a black *dishdasha* and black-and-white-checked *keffiyeh*, Abu Haider smiles broadly as he welcomes my partner Peter and me aboard his *mashuf*, a long, narrow canoe that is the traditional mode of transportation in the fabled Mesopotamian Marshes of Iraq. It's a bright April day in 2018, and he is taking us through Al-Hammar Marsh between the Tigris and Euphrates Rivers. Mesopotamia means "land between the rivers," and this wet region, once covering more than 5,800 square miles, is thought to be the Garden of Eden by Bible scholars. The rivers that supply water to the marshes originate in the mountains of Turkey, Iraq, and Iran, then tumble downhill. Here on the nearly flat Mesopotamian Plain, they slacken and spread in an apogee of slow water.

With us is Jassim Al-Asadi, director of local NGO Nature Iraq's southern region of operations. Al-Asadi has played a significant role in protecting these marshes. He grew up and still lives in our jumping off point, Chibayish, a town of around thirty-five thousand people that's a seventy-five-mile drive northwest from Basra. Al-Asadi recalls that, when he was a kid fifty years ago, the whole town rested on islands built of reeds, save one high street on which lay the school, the medical center, and other critical services. Everything else was water and canals. Although *mashufs* were traditionally propelled by long poles, modern ones also have motors. We chug down a canal and enter a water world.

Our slim boat shims through narrow passages of *Phragmites aus-*

FIGURE 3.1 On the Mesopotamian Marshes, people build homes out of *Phragmites australis* in an ancient style dating back nine thousand years.
Photo © Erica Gies

tralis, the giant grass towering twenty feet above us. Other water-loving plants nod in the light breeze: papyrus, cattails, sedges, and rushes. Swamp hens step through shallows on oversized feet built like snowshoes for mud. Purple herons and black-and-white pied kingfishers swoop by. The fishers rest up between hunts atop a bank of hardened mud in which they've excavated little apartments. In places, the grass thins and long vistas open across the water, punctuated by homes made of reeds in the ancient style—half-cylinders, open-side down. Some show signs of modern accoutrements, such as a plastic tarp, swamp cooler, or satellite dish.

The people who live here, the marsh dwellers, are descended from myriad backgrounds and speak various languages. Men wear pants and shirt or are dressed like Abu Haider. Women all wear *hijab* but in different styles, and some wear black *abayas* while others cover themselves in bright colors. But they share a common lifestyle and culture that began with the Neolithic fisher-gatherers who settled this area nine thousand years ago, traditions that grew and blossomed into the fabled Sumerian culture three thousand years later. In the mid-twentieth century, the population of marsh dwellers living atop the water was estimated to be between fifty thousand and a hundred thousand. This ancient chain of human life lived upon water is a reminder that people and wetlands are not fundamentally incompatible, that filling them with dirt is not required. Today, these wetlands still teem with other forms of life as well, serving as one of the world's most important stopovers for migratory birds and as breeding habitat for Persian Gulf fisheries.

But like the majority of wetlands and waterways around the world, they face relentless pressure from human development, including gray infrastructure. Many areas have even further diminished their major wetlands and waterways, including Florida's Everglades, Virginia's Great Dismal Swamp, and Flanders in Belgium, just to name a few. In those places, millions of people now live or farm on tenuously dry land. Although the transformation from wet to dry may have occurred decades or centuries ago, water continues to periodically mark its territory, an unbalanced gyration between flood and drought. As water detectives recognize this reality, they are trying to restore as much function as possible to waterlands within the constraints of human habitat. In Mesopotamia, restoration has largely been left to natural processes. In other places, such as Thornton Creek in Seattle, Washington (which we'll visit later in this chapter), reclaiming space for water is carefully managed by people. But either way, restoration is most effective when water's relationship with all the other elements in a system are allowed to flourish, from soil to microbes to larger critters, including insects, fish, and others.

Perhaps the reason the marsh dwellers have endured here for millennia is because they accept their water world for what it is—a rich source of life—rather than trying to drain it. In so doing, they have managed something remarkable: living on and harvesting from these wetlands without destroying them. Their approach is rooted in observation, tradition, and respect.

They embrace the bounty the marshes provide, starting with *Phragmites australis*, the fast-growing grass that is often viewed as an invasive scourge elsewhere. Here it's a native renewable resource

FIGURE 3.2 At the end of a day harvesting *Phragmites australis*, this woman's motor gave out, so she was poling home. We gave her a tow.
Photo © Erica Gies

used to make just about everything: houses, fences, animal pens, floor mats, cultural centers, and more. As our *mashuf* chugs along, we see people harvesting and processing *Phragmites*. Two young women dressed in colorful *abayas* are unloading stalks from their *mashuf*. Later, we pull up beside a woman in black, her *mashuf* piled high with today's harvest, poling home at dusk. Her motor had given out. We grab hold of the edge of her boat to give her a short tow. These communities ship grass products all over the Middle East. Trucks piled high with cut *Phragmites* or products such as woven mats or portable partitions head down the highway. Reeds bound into tight cylinders make the framework for stables or buildings, including the grand *mudhifs*, where community business and disputes are adjudicated by an elder. When laid on their side, the cylinders serve as backrests for people sitting on the floor, a typical practice in the region. The reed, when young and fresh, can even be food, as Al-Asadi demonstrates out on the marsh, pulling one up and unwrapping the root from its casing. It's crunchy, somewhere between

FIGURE 3.3 *Phragmites* reeds bound into tight cylinders make the framework for stables or buildings, including this grand *mudhif* in Chibayish, where community issues are adjudicated by an elder.
Photo © Erica Gies

jicama and bell pepper, but tastes bland. It doesn't strike me as a culinary sensation, but I imagine it's a welcome supplement to a diet that is largely animal protein.

Marsh dwellers fish with tridents, like so many Neptunes, and

trade with people in the drylands for vegetables. They also keep water buffalo for milk and cheese, which Al-Asadi serves us for breakfast during our stay, along with fresh herbs. The flavor of the cheese
is mild, a bit like mozzarella. The buffalo wander at will, and we see
them munching here and there, or swimming to a promising spot.
Big and black, with small, curved horns, they are docile like dogs,
seemingly members of the extended family. Our boat passes a tiny
girl standing next to one of her family's buffaloes, who is kneeling on
its front elbows to reach some hay.

Middle Eastern cultures are famous for hospitality, and I found
that to be true in the marshes. As we ply the waterways, people
wave from their homes and boats and call out welcomes. Passing
one homestead, Abu Haider and Al-Asadi greet the family, who
they know, and we are invited in. We step off the boat onto an island
built over generations by cutting and piling reeds and mud. Walking across its surface feels springy, like tundra, or a bounce house.
Entering the family home, we meet a woman and five small children. It's cool in the shade of the roof, and the building's walls are
arranged slightly askew to ease in the breeze. We sit on the floor
while Abu Haider smokes and chats with the woman in Arabic, and
the children excitedly play peek-a-boo with me and Peter. At the
opposite end of the perhaps 160-square-foot, single-room house is
the kitchen. It has a cylinder of gas for the stove and a tub for washing dishes. Aside from using *Phragmites*, people find much else they
need in their local environment. Outside, the family has a conical
bread oven fashioned from the marsh's sticky brown mud that hard-
ens into near concrete. It's fueled with sun-dried patties of buffalo
poop. On a fence beam, drying patties are each marked with our
host's handprint, lending a bit of flair.

Numerous legendary cultures have risen and fallen around the
marsh dwellers over millennia. Some of them converted parts of the
wetlands to irrigated agriculture. But the marsh dwellers hung on in
the remaining wetlands, retaining their customs. And as other civilizations passed into history, leaving their irrigation infrastructure to
rot, the rivers repeatedly retook their marshlands.

That's not to say that marsh dwellers' lives are all buffalo cheese and *Phragmites* snacks. Without sewage treatment, diseases associated with dirty drinking water have been common, although that's changing. Some modern practices, on the other hand, have eroded traditional stewardship, such as participation in a cash economy, which encourages the overharvesting of fish. But marsh dwellers also have an ethos of caretaking that differs from many Iraqis who live in cities, according to Suzanne Alwash, a geologist and cofounder of Nature Iraq. She relays these beliefs, collected through a survey, in her fascinating 2013 book *Eden Again*:

> They ... expressed an unexpectedly high respect for wildlife and the environment, strongly agreeing that animals and plants have the right to exist even though they may be of no use to mankind. They maintained a clear vision of a sustainable economy and environment, expressing the belief that it was right to sacrifice their current income and standard of living so that the next generation should benefit from preservation of the earth's natural environment.

The largest threats to sustainable life atop the wetlands remain external. In the early 1990s, during the regime of Saddam Hussein, Shiite rebels hid in the marshes. To destroy these hideouts, Saddam ordered the construction of massive diversion canals and dams that drained more than 90 percent of the original marshlands. After siphoning away the water, his minions poisoned and burned the area, leaving the wetlands a cracked, dusty expanse. According to a 2011 United Nations report, some 175,000 people were forced to flee. Many went to Baghdad or Babylon, refugees from Eden.

A Natural Restoration

Miraculously, the marshes survived that assault. After the 2003 US-led invasion of Iraq and Saddam's fall, many marsh dwellers returned from the cities and busted holes in the regime's levees. It was essentially a just-add-water approach to restoration, although wetlands experts from around the world advised.

Alwash writes:

> A consistent pattern emerged as each area was re-flooded. At first, the dry land was replaced by an open-water lake. Then wetland vegetation began to grow around its shoreline and on shallow banks within the lake. Eventually, the steady growth of reeds resulted in emergent marsh vegetation supplanting the open-water lake. None of the 3,400 square kilometers of new marsh vegetation required human intervention, no one planted seedlings or even reseeded at all—an astounding achievement for the world's largest (and most disorganized) wetlands restoration project.

Wetland plants were able to resprout because their roots and seeds remained in the soil. Many animals had hung on in nearby Al-Hawizeh Marsh, which straddles the Iraq–Iran border and, due to political considerations, was largely spared by Saddam. From Al-Hawizeh, animals were able to return to the reflooded marshes in Iraq: fish, endemic birds such as the Basra reed warbler, the Euphrates soft-shelled turtle, and a subspecies of otter thought to be extinct after the draining but common enough in the 1950s that local people kept them as pets. The fact that *Phragmites* is so hearty was a boon. Plus, although the ecosystem has been repeatedly disturbed, it is resilient, having been modified by human hands for millennia. The marsh dwellers cultivated the ecosystem by "tending their watery garden, harvesting its greenery, burning the dried reeds for rejuvenation, and herding their livestock to maintain its channels," writes Alwash.

Saddam's attack on the marshes had shrunk them to just 300 square miles in 2003. But by 2006, marshes and lakes once again covered 58 percent of their 1970s extent, or about 2,100 square miles. In their current form, the marshes are protecting surrounding cities and farms from flooding. Evaporation from the marshes also cools the local climate, and water's presence reduces desertification and dust storms.

But subsequent drought years have seen the marshes' expanse diminished. During my visit, I saw the limits of the restoration victory. One day, we drive from Chibayish to Nasiriyah to see the six-thousand-year-old Sumerian city of Ur. What I don't see, to the southwest of the Euphrates River, is wetlands. The road marks a boundary between two of the three main sections of the historical marsh. Today there isn't enough water to fill them all, so Al-Hammar Marsh is largely dry, having retreated to the distant and confined Al-Hammar Lake. We come upon a couple driving their large flock of sheep down the road, bandannas covering their noses and mouths from the dust. Behind them are houses and sheep corrals made of dried mud on bare land. Turning to the other side of the road, I see a woman and her young son walking along the edge of the Central Marshes with their buffaloes. Beyond this watery expanse, out of sight to the east of the Tigris River, lies the Al-Hawizeh Marsh, which extends into Iran. It is also struggling.

At Ur, which is mentioned in the Bible, we climb up its ziggurat. From the top of the structure's bitumen-mortared brickworks, I can see an American-built prison that houses the most dangerous ISIS captives. Ur is part of a UNESCO heritage site commemorating the Sumerian civilization, including the marshes. But due to decades of bloodshed in the region, there are no tourists. Today the only other visitors are a French documentary film crew and an Iraqi family from Chibayish. The site's wonders include the world's oldest standing arch; inscriptions in cuneiform, the world's oldest documented writing; and countless pots and other artifacts poking out of heaped mountains of material waiting to be sifted by archaeologists. Also scattered everywhere are shells. In the Bible, Ur was a port city at the mouth of the Euphrates. A major flood two thousand years ago caused the river to avulse to a new channel, leaving Ur stranded and dry.

Water levels in the marshes have always fluctuated due to seasonal changes, drought, and wet periods. Climate change is now exacerbating the dry swings. But the primary cause of shrinking acreage is upstream dams and water diversions—especially Turkey's massive Great Anatolia Project to turn its arid east into an agricultural heartland. Before Turkey's Ataturk Dam was completed in 1990, the average flow on the Euphrates where it enters Iraq was more than thirty-five thousand cubic feet per second. In the '90s, it decreased to about twenty-one thousand cfs. And in April 2009, a high-flow time of the year, it had fallen to around eight thousand cfs. Since then, Turkey has filled additional reservoirs upstream, including the contentious Ilisu Dam on the Tigris River that flooded the twelve-thousand-year-old town of Hasankeyf, Turkey, in 2020. The Tigris River flowed around twenty-one thousand cfs before the Ilisu Dam was built; after, flows dropped to eleven thousand cfs.

Those decreases have also harmed water quality because it has concentrated the impact of human- and buffalo-caused pollution on the marsh. When some of the water connections sever due to low levels, waste doesn't get flushed out and oxygen in the water decreases. Iran's dams and diversions are also contributing to the dramatic decrease in fresh water flowing downstream, causing negative impacts beyond the marshes. It's allowing salt water to push further upstream from the Persian Gulf and through the city of Basra, sparking riots and militia conflicts over increasingly salty and unusable drinking water.

In 2013, government officials made the Mesopotamian Marshlands Iraq's first national park, and in 2016, UNESCO bestowed upon it World Heritage status. But it's unclear if these designations will be enough to save the rich wetlands and the marsh dwellers' ancient way of life. Turkey has long scorned international water-sharing agreements. And the dam building continues.

Dam Damage

Dams might seem to be a way to help water reclaim more of its slow periods on land. But often they are too industrial and too big to deliver what water and aquatic life wants. Behind a dam, water stalls artificially deep and long. It warms in the sun, harming fish, and is discharged according to people's timeline for water or electricity—not nature's cycles of seasonal highs and lows that many species rely on for different parts of their lifecycles. Water leaving a reservoir is often moving faster, with greater pressure, than water entering the reservoir. These deviations from water's natural flows cause problems in river systems that affect people too. While dams vary in size from microbarriers in local streams to massive river stoppers, the biggest problems are caused by the latter, those epic monuments to control. And there are now a lot of "big dams," defined as fifty feet high or taller: about fifty-eight thousand worldwide.

The first megadam I saw was Hoover, spanning the Arizona–Nevada border. At eight years old, on a road trip with my grandparents, I was awed by its smooth, 726-foot-high face and especially by the vast lake behind it, a beacon of life in a sere landscape. You could walk or drive across the top of the barrier, and art deco clocks mounted on pillars showed the hour time change between the two states. Both in my grandparents' estimation and on informational signs, there was no question that this dam was Progress. It was turning a "worthless" desert into an agricultural wonderland and supporting the growth of cities such as Phoenix and Las Vegas. In other parts of the world, the allure of dams is electricity and the industries it can power.

Yet, as the world's first megadams have aged, people have witnessed more of their pitfalls.

Dams block fish passage. Hoover stymied the Colorado River's distinctive razorback suckers, who are now on the brink of extinction. Fish loss is particularly devastating to subsistence fishers who rely on them for their main source of protein, such as the sixty million people living along the Mekong River in China, Laos, Cambodia, Thailand, Myanmar, and Vietnam. There can be untold knock-on consequences. Consider freshwater mussels, many of which port their larvae to new streams reaches on the backs of fish. Dams blocking fish are a primary cause of the extinction of thirty North American species of freshwater mussel over the last century and imperil 65 percent of those remaining. In turn, the loss of mussels—because they are filter feeders who keep streams clean—has been a factor in the decline of bottom-dwelling plants. And less waste from mussels is dropping to the streambed, where it once fed algae, insects, and other invertebrates that in turn feed fish.

As with Turkey and Iraq, dams often create human water haves and have-nots. They give the illusion of increasing water supply, but they are actually water grabs. Between 1971 and 2010, 20 percent of the global population gained water from human interventions on rivers, including dams, but 24 percent were left with less water, according to a 2017 study. Hoover is part of this pattern. Dams and diversions on the Colorado River in the western United States denied water to farmers downstream in Mexico—until a diplomatic breakthrough in 2021. Dams also harm downstream farms by starving delta croplands of sediment, reducing soil nutrients over time and causing land loss. Dams push people off their ancestral lands, flood irreplaceable cultural sites like those in Hasankeyf, and destroy habitats for plants and animals. In all these ways, dams are an environmental justice issue.

Dam boosters justify some projects by saying they bring electricity to people who don't have it. But often, power generated is shipped to cities or sold to nearby richer countries, the profits lining the pockets of the elite. The recently completed Xayaburi Dam on a Mekong tributary in Laos is supplying electricity to Thailand; the controversial Inga 3 dam planned for the Democratic Republic of Congo would primarily power mining companies and South Africa, rather than serve the 91 percent of Congolese who lack access to electricity. And while people sometimes knock wind and solar because they're not steady sources of electricity, climate- and human-caused changes in water supply are also making hydropower increasingly unreliable. A low river flow can cut hydropower dramatically, as happened for months in southern Africa following the 2015–16 drought. At that time, Zambia relied on hydropower for 96 percent of its electricity. Drought and energy shortages plunged it into an economic crisis.

Despite these negative impacts, the dam-building boom continues, in part because hydropower is seen as a solution to climate change—a carbon-free source of power that doesn't require burning fossil fuels. But hydropower's reputation for low emissions has come under scientific scrutiny in recent years. Cement, used to build dams and other infrastructure, is energy intensive to formulate; its production accounts for 8 percent of global carbon emissions. The reservoirs of deep water that dams create flood vegetation, which releases methane as it decays. When water levels fall again, soil is exposed and dries out, releasing stored carbon. An article in the journal *Nature Climate Change* concluded that "emissions from tropical hydropower are often underestimated and can exceed those of fossil fuel for decades." That means when hydropower replaces coal or natural gas-fired power, it can take decades to break even on hydropower's embedded carbon—a liability that is often ignored during planning or poorly addressed. A 2021 study in the journal *Land Use Policy* looked at a requirement to offset carbon emissions from Himalayan dams that flooded forests. Typically, people installed plantations, leading to biodiversity loss, "abysmally low presence of surviving saplings (up to 10 percent), interspecies conflict, infringement on local land usage, and damage by wildfires and landslides."

Dams can also harm those people they purportedly benefit. An emerging field called sociohydrology is generating surprising insights into gray infrastructure such as dams. Until recently, water engineers would only calculate "natural" flows, not taking into account the effect humans have on water supplies. Now that they are realizing the scale of our impacts, they are beginning to tally them.

For example, although building a dam is often seen as a "new" source of water, in fact, dams tend to create new *demand* for water, according to Giuliano Di Baldassarre a professor of hydrology at Uppsala University in Sweden. In response to water scarcity, managers build new reservoirs to store more water, which enables more water consumption, recreating the shortage. Las Vegas is a textbook case. Decades ago, the population was projected to reach 400,000 by 2000, so the city built a pipeline to tap into Lake Mead, the reservoir created by Hoover Dam. Then its population ballooned to nearly 1.6 million, increasing water demand. There are many similar examples: Phoenix, Los Angeles, San Diego. Projected water shortages on the coast of Spain in the 1980s were "solved" by building a water pipeline from another region. That fed a frenzy of condo building; now those coastal provinces again need more water. (Some conservation and reuse programs aim to override this trend. In water management circles, conservation is considered a source of "new" water.)

A related phenomenon, called the reservoir effect, is the false sense of security that dams give people about the quantity of water available. The big pool of water sitting there masks a drought, reducing incentives to conserve or adapt until shortages get dire. The reckoning, when it comes, is particularly harsh. These paradoxes show the benefit of using local water: people in a region learn to live within their water means.

Confronted with all these downsides, some people in the United States are now removing dams, often motivated by the hope of restoring fish populations. Sometimes there's little to lose because the reservoirs have filled up with silt and become useless. Nevertheless, for decades, international development lenders such as the World Bank and the International Monetary Fund pushed countries seeking investment to take the western development path of gray infrastructure, at times replacing traditional methods that were more sustainable. The water scientist and Shinnecock citizen Kelsey Leonard calls this hydrocolonialism, a subjugation of both nature and ourselves. Indeed, many countries have internalized the notion that dams are harbingers of prosperity. Currency bearing images of dams can be found from Mozambique to Laos, Kyrgyzstan to Sri Lanka. As international lenders have at last withdrawn some support for big dams, regional and national development banks have stepped up to fund them in Latin America, Asia, China, and Turkey.

Dam building continues apace in Canada and Norway, and a boom is underway in many lower-income countries, with more than 3,700 hydropower projects planned or under construction on the world's rivers, especially in South America and Asia. But as rivers get sicker, some people are realizing that the costs outweigh the benefits. Cambodia, where tens of millions of people rely on subsistence fishing that will be devastated by new dams on the Mekong, recently put a ten-year hiatus on its planned dams.

Systems Synchronicity

The many problems caused by dams and other human interventions on natural waterways—levees, diversions, straightening, deepening, drying and more—are indicative of the complexity of natural systems. An ecologist I interviewed many years ago told me, "It's not rocket science. It's way more complicated than that." Scientific discoveries about natural systems can change our whole perception of them. Consider forest scientist Suzanne Simard's 1997 discovery that miles of underground fungal filaments—called mycorrhizal networks—act like trading and communications systems for trees. Through the networks, they share food, water, and news about invading pests. When a tree is felled and can no longer photosynthesize food, other trees may even keep its stump alive.

Rivers and wetlands are similarly complex, full of life connections at myriad scales. Water interacts with the land around it, moving over and through it. Patterns of water flow, sediment dispersal, and channels shape and are shaped by aquatic and riverbank plants and animals. Healthy river and wetland ecosystems maintain themselves through these relationships. That complexity is why it's difficult to fully restore natural systems when we damage them.

It's not rocket science; it's systems theory. Systems theory recognizes these interconnections and tries to understand their interwoven relationships. An undisturbed ecosystem is resilient because it can absorb some level of external disruption by adapting. There's flux within the system, and it can even demonstrate emergent behaviors in response. At the same time, those links among elements in the system mean that the destruction of a single element or partial damage to several elements has the potential to tip the entire system into dysfunction.

Our incomplete grasp of complexity is why our interventions on waterways have had such negative consequences. When we need energy, or water, or want to stop flooding, we attempt to address these needs single-mindedly. For example, when engineers plan a dam, they look for a narrow canyon to wedge the barrier into, with space behind it that they can use for water storage. If they want to generate electricity, they look for a big elevation drop. But if those are the criteria, what else are we ignoring—and sacrificing?

The scientific method usually divides a system into components to eliminate variables in order to study a piece of it. While this approach can provide clear answers to a particular question, it can also obscure the big picture. Now additional problems such as climate change are pushing scientists to factor in whole systems and work with people in other disciplines.

That approach has something in common with people who live closer to the land, such as the marsh dwellers. They have invested time in close observation and have a greater awareness of the intricacies of natural systems, although they may explain them differently than scientists do. Another example is the way some Indigenous people in northwest North America categorize things they encounter in the environment. Rather than dividing critters into categories such as insect, bird, mammal, they classify them using kinship clusters, a group of species who rely on each other to survive. That type of classification contains rich, useful information about their relationships. Similarly, Indigenous languages worldwide often contain detailed information about local ecology and systems embedded in, say, names for plants or stretches of creek.

*

The complexity and variation of types of wetlands around the world is another window into water's multifaceted nature, as it interacts with different soils, plants, and climate. Wetlands are places where water stalls on the land at least part of the year, or where the water table is high enough to support plants who like wet roots.

There are four main kinds: marsh, swamp, bog, and fen. Which name applies depends on whether they lie by a river, lake, or ocean or are isolated from other water; the soil and geology; whether they're fed by rain, surface water, or groundwater; and what types of dissolved matter they contain that influence the water chemistry. All these factors affect which plants can grow, creating different ecosystems.

Marshes get most of their water from the surface and a bit from groundwater. The water usually has a neutral chemistry, not too acidic or basic, making it homey for many plants and animals. Freshwater marshes include wet meadows, prairie potholes, vernal or ephemeral pools, floodplains, dambos, and playa lakes. Tidal marshes, where salt water often mixes with fresh, include sloughs, salt marshes, mudflats, and bayous.

Swamps have wet soils or standing water at different times of the year. They are dominated by woody plants, such as the bottom-land hardwood forests of cypress and tupelo in the US Southeast, or mangrove forests in Asia, Central America, and elsewhere. Their organic soils are nutrient rich and also support a lot of plants and animals. For that reason, both swamps and marshes, when healthy, make good foraging and hunting grounds for people.

Bogs, sometimes called mires, have spongy peat, a type of dirt that is primarily decayed vegetation rather than mineral components like rocks. Peat can be many feet deep, built up over centuries. Bogs get all of their water from precipitation, rather than groundwater or streams, so they are acidic, with low mineral content and nutrients. For that reason, only specialized plants such as cranberry, blueberry, and carnivorous plants can thrive. Fens also form peat but fill with runoff and groundwater. They are less acidic than bogs and support more diverse plants, such as grasses, sedges, rushes, and wildflowers.

When wetlands are destroyed, it's not just these complex ecosystems and water storage that are lost. Healthy systems process nutrients, including sulfur, phosphorus, and nitrogen. Wetlands also store atmospheric carbon dioxide captured via photosynthesis in their decomposed plants, especially those that stay wet the longest. That's why peat is a superstorer of carbon. Although just 3 percent of Earth's land is covered in peat, it holds 30 percent of land-based carbon dioxide.

Some of the largest remaining wetlands in the world—all under threat—include the Amazon basin (swamps, mangroves, and more); the West Siberian Plain (northern peatbogs), the Pantanal in South America and the Okavango delta in Southern Africa (seasonally flooded grasslands), and the Sundarbans in Bangladesh (tidal mangroves). The Mesopotamian Marshes, which have dwindled to less than half their historical reach, include freshwater floodplains and river deltas, ephemeral pools that come and go with the seasons, and estuaries where the deltas meet salt water.

The Secret Life under Rivers

As shown by the long history of human civilizations ebbing and flowing across the Mesopotamian Marshes, water reclaims its space whenever given the opportunity. Busting open the levees to let water flow through again went a good distance toward restoring the marshes after Saddam's attempt to drain them. But in a densely built city like Seattle, where fewer healthy connections to the watershed remain, restoring waterlands to health can be trickier.

In north Seattle on a sunny September day in 2019, I park at the Meadowbrook Community Center near two cutting-edge urban creek restoration projects. They were needed to correct human errors that had led to repeated flooding of homes, roads, a high school, and the community center. Here, as in many now-urban areas, people logged the native forest up to the edge of the stream, removing trees that had caught and slowed rain runoff and reduced erosion by anchoring dirt with their roots. Attempting to solve these problems caused by deforestation, people straightened the creek to move water off the land quickly and armored the banks with concrete. Then they built houses on the denuded floodplains.

This pattern of urban development has been repeated worldwide, with some waterways further suppressed by encasing them in pipes and burying them. The result is what ecologists call "urban stream syndrome." The ailment is characterized by "flashy" flows (meaning large quantities of water quickly running off pavement and into streams), unstable banks, reduced biodiversity, and high levels of contamination with nutrients and other pollutants.

Human attempts to control water systems—including those upstream of Iraq and these in Seattle—have interfered with natural slow water phases such as wetlands and floodplains. What water wants, in these places, is to slow down, and in that slowness, water forges its complex relationship with life in local ecosystems. Because our gray infrastructure thwarts slow water, it bears much responsibility for that steep decline in fish, amphibian, reptile, mammal, and bird populations in and around lakes and rivers described in chapter 1. But it also causes problems for us, including flooding and local water scarcity. To repair these harms, we need to move beyond the control mindset and instead employ systems thinking to manage water holistically.

The Seattle projects set out to do just that, taking a novel approach to stream restoration that supports tiny forms of life and, in turn, recruits their help. The water detectives restored two stretches of Thornton Creek totaling 1,600 feet and covering about four acres. The result is dramatic: on the day of my visit, I enter a small urban oasis.

I walk down a pathway to the Thornton Creek confluence, where the North and South Forks meet, and Meadowbrook Pond. It's a lovely respite from the city: the water is clear; crows holler; squirrels race up willows, cottonwoods, and cedars. And these aboveground signs of life are just the proverbial tip of the iceberg. What makes the Thornton Creek restoration special is the water detectives' attention to a mysterious area *under* the stream. Similar to the discovery of the mycorrhizal networks under forests, stream ecologists are learning more about what goes on in something called the hyporheic zone (Greek, *hypo*, "under," and *rheos*, "flow").

Just below the streambed, the sediment and soil are filled with water. But this is not an underground aquifer—it's a limbo zone that's dynamic and contested, where groundwater pushes up into surface water and the stream forces water back down underground. Like the river on the surface above it, water within the hyporheic is also flowing downstream but orders of magnitude more slowly. In a healthy system, that underground river can be vast and hold a huge volume of water. The expanse and depth of the hyporheic varies with topography, gradient, sediment size, channel shape, and quantity of flow. Along a large river, the hyporheic zone can extend underground laterally more than a mile from the banks and reach below the bed a hundred feet or more—although the main hyporheic flows are shallower, in the top three to ten feet or so. Time spent underground modulates water's temperature, cooling streams during the hot summer months and warming them in the cold winter, creating a more stable environment for the fish who live there.

Within the hyporheic, water chemistry and life forms differ

from those in the adjoining groundwater or surface water. Such in-between zones are called ecotones, liminal spaces known for high biodiversity because their neighbors' species mingle there along with unique ecotone denizens. Hyporheic zones are home to microbiomes that, like our own guts' microbial diversity, are a determinant of wellness.

Thanks to slow water, critical physical, chemical, and biological processes happen here, including soil aeration, water oxygenation, pollution cleanup, waste decomposition, and food creation. These services have earned the hyporheic zone nicknames, such as the "liver of the river." Yet despite its importance, the recent Thornton Creek projects were among the world's first to physically recreate a hyporheic zone in an urban stream—and to inoculate the sterile substrate with tiny life forms harvested from a nearby healthier stream.

The Shape of Slow Water

The repair of Thornton Creek began in the 1990s when Seattle Parks and Recreation department began buying out homeowners and removing houses that flooded repeatedly to give the creek the room it steadfastly refused to relinquish. City officials also hoped to reduce stormwater runoff and to sustain remnant salmon populations. Thornton traverses fifteen miles, draining an 11.4-square-mile watershed that's home to seventy thousand people. The two restoration projects' 1,600-foot scope might seem too small to make a difference. But for people who knew the creek's historical shape, the routine flooding at the Kingfisher and Confluence project sites was predictable; these spots were originally floodplains. Experts thought that handing them back to the creek could absorb significant water.

The projects' champion was Katherine Lynch, a biologist working for Seattle Public Utilities. Earlier in her career, she'd worked on wilder, rural streams. When she was first assigned to urban creeks, she balked: "I thought, 'Excuse me: I work on river systems. And now you want me to work on ditches? No thank you.'" She eventually realized that her resistance was rooted in sadness, a feeling that maybe urban creeks were beyond salvaging. Thornton Creek had a reputation as one of the worst. But early on, as she explored the riprapped channel, she found small signs of life, giving her hope. She also met human residents who called it "our creek," and she realized that local people loved it. She resolved to make the most out of the Thornton projects.

After visiting the site, I speak by phone with Mike Hrachovec, an engineer with Natural Systems Design, the firm that devised the projects. He explains to me that when people straighten a meandering creek into a narrow channel, the water's energy is funneled, speeding its pace. Additionally, in a city, rainfall often can't soak into soil where it falls but instead runs off sprawling pavement and rooftops, creating much higher peak flows than in a natural landscape. Together these impacts create what Hrachovec calls a "firehose" effect. The firehose sprays water down the channel "ten to one hundred times faster than historic rate," cutting down into the earth in a process called incision. This is classic "fast water," and it means that most of it doesn't have time to soak into the ground in the channel or floodplain, significantly decreasing the amount of water locally available during dry periods. Historically, "streams had a much more consistent flow range between winter and summer," Hrachovec says. Healthier hydrology could help supply dry-season water in drought-stricken areas.

Blocking streams from floodplains also interferes with natural processes that shape and slow flow. The action of creeks and rivers periodically overflowing onto surrounding lowlands dissipates water's energy, reducing erosion. Slow water also deposits silt, helping to build the floodplains. And as water meanders across its floodplain, then returns to the river, it shapes channels over time to better accommodate high flows.

These pulses of slow water also allow the river and the floodplain to exchange nutrients and sediment, making them both more productive while cleaning the water. As fish biologists in California's Central Valley discovered, many fish spawn on floodplains, and hatchlings fatten up on the algae and plankton that grow in shallow, warm water. Because these plants and critters evolved under a pattern of small, predictable floods, a floodplain cannot function ecologically without being connected to the river and without regular flooding.

When we disrupt these physical processes, high water has nowhere to go—except to flood human homes and businesses. In response, people here built berms to keep the water in Thornton's deepening channel. That further strengthened the firehose effect, and water cut down even deeper through the soft, absorptive material forming the hyporheic zone. It scoured away that sediment, gravel, wood debris, and nutrients until it reached bedrock, removing the hyporheic zone almost entirely.

Over the last couple of decades, stream ecologists worldwide have tried to repair damaged creeks by restoring their shape, from straight, deep channels to more natural sinuosity, a long, messy curl with strands separating and rejoining. To entice fish, they've plopped wood into the channels to create more habitat. Seattle Public Utilities planned to follow this approach and also wanted to restore Thornton Creek's connection to its floodplain, a first for Seattle but a tactic that is becoming more common today in restoration.

Aside from reducing flooding by giving water more space to slow, the idea is to build homes for myriad species by creating a mosaic of habitats—gravel bars, meander belts, side channels—with each area supporting different critters. A more natural stream might also have oxbows, riffles, or wetlands. The diverse topography also slows water, and added vegetation growing alongside streams provides food, places for birds and insects to stand and hunt, openings in soil for water to soak into, and other important life ingredients, such as speckled sunlight and leaf litter.

Next-Level Restoration

It all sounds very enticing. The practice of restoring a stream's curves and reconnecting it with its floodplain is based on an assumption that by recreating some of the physical parts of a healthy stream, plants and animals will repopulate it. Ecologists call this the "field of dreams" approach, after the classic 1989 baseball film by that title with the famous line "If you build it, they will come." The assumption has been borne out—to a degree. When people restore a damaged system, life returns, as we saw in the Mesopotamian Marshes. But in the last decade or so, ecologists have been noticing that, in many cases, the life isn't very diverse. A lot of species are missing.

A landmark 2019 paper in the journal *River Research and Applications* put it boldly: "River management based solely on physical science has proven to be unsustainable and unsuccessful, evidenced by the fact that the problems this approach intended to solve (e.g., flood hazards, water scarcity, and channel instability) have not been solved."

Without much of the full biodiversity native to a stream, it loses capacity to soften floods and water scarcity, provide habitat for plants and animals, clear the water of pollution, and store carbon dioxide. It also cannot maintain itself without ongoing "help" from humans that can be costly, fleeting, or misguided because people don't fully understand all facets of the system. The call-to-arms paper exhorts scientists to use "the power of biology to influence river processes," something the authors call biomic river restoration.

In another example of the "shifting baselines" syndrome, most of us today have little idea of how rich in life a healthy river system actually is. In many places in the world, we are used to concrete channels, bare streambanks, cropped grass, or soil compacted by livestock. But in a more natural state, a river is lined with water-loving plants whose roots tap shallow groundwater. Often these are tall trees or reeds, forming a characteristic green stripe, even in arid places, called a riparian corridor. Those plants help hold the stream bank from erosion and provide habitat for wildlife. Ecosystems around the world differ, but all life comes to the river.

I didn't realize the extent to which every place I'd seen was incredibly diminished until I visited the small South American country of Guyana, where 85 percent of the land is forested, much of it with primary ecosystems. Guyana is derived from a Wai-Wai word that means "water people," and its small population clusters largely along the Caribbean Coast. The interior Amazon jungle and savanna remains home to several thousand Indigenous people—Macushi, Wapishana, Patomona—many of whom still live in small villages accessible primarily by rivers. In 2009, traveling by small boat along the Rupununi River, I saw five-foot-tall Jabiru storks, flocks of rainbow-colored macaws, potoos standing stock-still atop dead trees to blend in with the wood, black skimmers (who were surely the inspiration for Gary Larson's cartoon birds) standing in groups on sandbars. Fifteen-foot-long caimans drifted across the water, only their eyes and back showing; snakes hung in trees; six-foot-long giant river otters popped their heads above the surface in a move called periscoping, looking for danger, screaming out to each other in an ear-shattering cry. Within the river were eight-foot-long arapaima, a major food fish for the locals; river stingrays; and piranhas. Capybaras and agoutis came to the bank to drink, along with jaguars, and capuchin monkeys chased each other through the treetops. One night, the air was so thick with moths that I had to put a bandanna over my mouth to avoid inhaling them.

This picture may sound exotic—something possible only in the rainforest. And while it's true these critters are specific to the jungle, rivers in temperate lands were likewise incredibly rich places before industrial agriculture and modern development. Until not too long ago, China had plenty of ten-foot-long giant catfish and pink dolphins. The Mississippi River was populated with six-foot-long alligator gar, a fish who coexisted with dinosaurs. According to an account of a mid-nineteenth-century traveler, San Jose Creek off San Francisco Bay was then so rich with thousands of waterfowl, "when we wanted to carry ducks and geese to the ship . . . we simply threw clubs at them, and knocked them over, and thus saved our powder and shot."

What Lies Beneath

With that historic richness in mind, it's easier to understand why managers of restored urban creeks have found that many sites need constant upkeep; they're still so far from their natural state, missing space and life for critical processes. Lynch thought a more complete physical and biological system might maintain itself better, requiring less human intervention. On a 2004 trip into the forest, she was amazed when a colleague pointed out that they were standing atop a hyporheic zone. "I'm looking around at trees and ferns, thinking, how is that possible?" She began to ponder the fact that urban streams were missing not just horizontal space to meander and migrate but also the thicker substrate on the streambed that she saw in wilder streams. She began tracking a growing body of scientific knowledge about the hyporheic zone. Then, in 2009, a compendium of research called *The Hyporheic Handbook* was published in the United Kingdom and "became a bible for me," she says.

She realized that the hyporheic zone was an overlooked missing link in urban stream restoration. Even people considering horizontal space for water in floodplains weren't thinking about the vertical play. When the hyporheic zone is washed away or depopulated, pollutants and waste can build up and larger creatures can go hungry. It's akin to when humans have digestive issues because their gut fauna is deficient. But as far as she could tell, no one had taken the next step: to build from scratch a missing hyporheic zone, then repopulate the sterile area with microbial life.

Lynch wanted to go for it. But urban stream restorations have big price tags, tight budgets, and high stakes: ensuring peoples' properties don't flood. Convincing any agency to stick its neck out to try something radically new is hard. Plus, there was a more basic hurdle: "People had no idea what I was talking about." So when budget concerns arose, one decision maker quickly axed the hyporheic element. In response, Lynch penciled out the cost and found that piece was just $300,000 out of the two sites' combined $10.5 million budget. She also argued that if the restorationists could repair a stream's natural systems, the creek might use and move sediment on its own, ultimately saving money by reducing the need to spend an average of $1 million a year dredging a nearby stormwater pond. Ultimately Lynch prevailed, in part by promising to incorporate monitoring into the design so scientists could study whether it worked and gather data to inform subsequent projects.

Paul Bakke, a hydrologist focused on geomorphology, did baseline measurements, which clearly showed that Thornton Creek's hyporheic zone had been almost completely scraped away. For Bakke, the projects were personal. He'd grown up in the 1960s and '70s along Thornton Creek, fishing for cutthroat trout and playing with water skeeters. Then, just before he entered high school, condos were built right along the creek's edge, cutting off his access. "These old haunts that I really loved, that were my sort of wilderness, really . . . were suddenly not just blocked but being paved over," he tells me via phone. "It was very upsetting."

Lynch tapped Bakke to team up with Hrachovec to design the project, thinking the geomorphologist and the engineer would bring together two very different perspectives. She wasn't wrong. Their partnership was rife with intense disagreements. One of the biggest battles? Bakke wanted to put larger gravel on the streambed to allow water to move more easily into the hyporheic. Otherwise, he feared, superfine urban dust washing into the creek could "plug up" the downward flow. But Hrachovec worried that large gravel might allow too much water to move underground, drying out the surface stream in summer and killing fish. To sort it out, they ran tests on computer models and in a large sandbox, modeling stream dynamics and trying different aggregates, meanders, and wood placement.

Satisfied at last, in summer 2014, they put the bulldozers to work. Hrachovec and his team removed the berms and dug out an area much wider than the stream. Because the firehose effect had scoured out the gravel and subsoil where the hyporheic zone used to lie, Hrachovec and his team created space for it too. To address urbanites' anxiety about flooding, instead of bringing Thornton Creek up to its former floodplain, they created a new floodplain at a lower level for the creek to occupy during storm flows. Then they added nearly eight feet of sediment and gravel to fill in some of the deeply incised channel so water in the creek could more easily reach its new floodplain and so hyporheic creatures could find a home.

On the day of my visit, I look toward the confluence of the North and South Forks of Thornton Creek and see a series of installations called logjams—basically, short logs crisscrossing the streambed. The logjams create tiny waterfalls, plunge pools, and pockets of nearly still water, dissipating the stream's energy away from eroding the bank. They also force water down into the newly created hyporheic zone and up out of it, increasing circulation. Other logs the team buried in the gravel of the hyporheic, creating another passageway between the zones. Over time, gravel falls out of the slowed water and builds up behind the barriers, creating a gentle grade. In the new floodplain, Hrachovec and his crew also added big chunks of trees and piles of woody debris to slow water when it spills over so it drops sediment, rebuilds the floodplain, and directs water into multiple channels.

The monitoring the utility required allowed Bakke and Hrachovec to track the water flow by temperature and tracers to confirm it was moving into the hyporheic zone. Result! They found water mixing at eighty-nine times the preconstruction rate. Seeing the stream function as nature, Bakke, and Hrachovec intended, the two designers were visibly relieved, shrugging off the months of anxiety. "They were high-fiving, laughing together," Lynch recounts fondly. But was that flow also supporting life?

The Life Hyporheic

In a healthy stream, there's a lot going on. Carnivorous fish eat macroinvertebrates such as crustaceans, worms, and aquatic insects, including dragonflies, true flies, beetles, caddisflies, stoneflies, and mayflies. In the interstitial zone between the rocks on the stream bottom live creatures smaller than macroinvertebrates called meiofauna, less than a millimeter long. Their group includes such characters as nematodes, copepods (sort of like tiny shrimp), rotifers, and tardigrades (a.k.a. water bears).

Below that is the hyporheic zone. Some macroinvertebrates move between the hyporheic zone and surface during their life cycles. The saturated sediment is a combination of mineral (rocks and sand) and biotic (living) material. Life forms include decomposing plants, tiny animals who break down matter and cycle nutrients, and predators who feed on them, including worms, springtails, and mites. The mineral aspects of the soil affect how water moves through it:

remember sand, clay, or gravel. But the biotic aspects also affect water flow. The critters dig tiny tunnels for themselves, keeping the soil aerated, allowing water to infiltrate. In turn, that moving water supports life, including salmon spawn. Adults lay eggs in the upper reaches of the hyporheic, in places where upwelling delivers higher oxygen levels to the developing fry.

Then there are the microbes, tiny ecosystem engineers. They metabolize inorganic compounds, converting them into food for plants and bugs. They feed other organisms who live underground where there is no light for photosynthesis. They move organic matter and nutrients between the hyporheic zone and surface sediments and play a pivotal role in the nitrogen, phosphorus, and carbon cycles. Anecdotally, they seem to be able to break down some pollutants such as petroleum leaks as well.

It's a whole little universe going on down there. The services that microbes perform are an outcome of their own agendas, including warfare. For example, "the antibiotics we take are weapons manufactured by microbes," says Linda Rhodes, a microbiologist with the National Oceanic and Atmospheric Administration (NOAA). The microbes are fighting each other in a competition for space. Microbes also help each other, such as one species making a product that others use, including biofilm (what makes stream rocks slippery when you step on them) and nutrients that help sustain the community.

Bugging the Creek

Kate Macneale, an environmental scientist for King County, where Seattle lies, has been thinking about overlooked life in rivers for decades. She monitors streams for insects as a measure of health, calling this the "bug score," which she explains to me with charming enthusiasm. Caddisflies, stoneflies, and mayflies are Macneale's star indicators because they are sensitive to changes. She has found a clear correlation between urbanization and lower bug scores.

Some harms are easy to diagnose. Stoneflies eat leaves they shred from streamside trees; no trees, no stoneflies. Many insects need to hide in the spaces between rocks; in urban streams, there is often too much fine sediment filling in those protective crevices. When stormwater rushes over streets, picking up dust from pavement and flooding the stream, those fast flows and fine sediment scour nutritious algae off rocks, wash out leaves, and turn the stream into a bug blender. Urban runoff poisons bugs with pollutants such as the ubiquitous dust from vehicle brake pads. "Bugs are particular about the conditions," says Macneale. "Some just can't hack it." Insects and microbes who can—the crows, rats, and pigeons of these ecosystems—multiply to fill the void.

Even if a "crow" insect fills a particular niche, losing a more sensitive species may cause a cascade of ecosystem failure. For example, some fish have refined preferences about the insects they eat. "I think of it like us in a salad bar," Macneale explains. "We all have our favorite things we eat. If you look at the diets of fish in urban areas, they're just not getting that diversity." Macneale thinks of the insects who can survive in urban areas as the french fries of fish diets. She admits, "I'm biased. I call them crappy bugs." And it's not just fish who suffer when some species are missing. Because different insects process nutrients in different ways, rich biodiversity brings many other ecological benefits, making the whole community more resilient.

Although the human instinct to solve this problem might be to just reintroduce missing species to a restored stream, it's not that simple. Ecologists are painfully aware of cautionary tales, such as someone looking to curb a species by adding a predator—who ended up eating everyone, not just the target animal. Even reintroducing native species can shake up a system that has adjusted to its absence, or inadvertently bring along pathogens.

While Macneale was wrestling with this ethical dilemma, vandalism intervened. She had set up an experiment on Longfellow Creek in Seattle to study bug behavior, using insects she'd collected from nearby healthier streams that she presumed were once native—although with urbanization, they had long been absent. She placed them in experimental stream channels that were supposed to contain all the bugs, so they could be collected and analyzed after the experiment. All was going swimmingly until the vandals trashed the site, dumping the bugs *Free Willy*–style into the creek. Focused as she was on restoring the experiment, this time with security cameras, Macneale didn't think too much about the bugs that got away.

Two years later, when she was sampling fish in the same creek, she saw a caddisfly in a fish's gut. "I didn't believe it. Then it dawned on me: we weren't seeing the individual we'd added [by mistake]; we were seeing the grandkid of that individual." She was shocked that this sensitive insect had not just survived in the degraded urban stream but had reproduced. "And it gave me that light bulb": maybe some insects were missing from urban creeks not because they couldn't survive the conditions, but because there was no nearby source of colonists. That could be one explanation for why the "field of dreams" strategy wasn't working.

A lot of assumptions that bugs would return quickly are based on studies where upstream habitat is healthy. "But we don't have that when the upper watershed is a Home Depot parking lot," Macneale realized. That's not hyperbole—the headwaters of Longfellow Creek "is literally a Home Depot parking lot." Especially for insect species who can't fly, "it's just really hard to get places," she says. Even for those who can, their range is only about 1.2 miles. "In urban areas, they would need to fly 10 kilometers [6.2 miles] to repopulate a stream because so many urban streams are so sick or buried in pipes."

She began to think seriously about taking the risk to reintroduce key bugs who could improve stream function. "If we're ever going to expect them to have a chance of recolonizing, we may need to help them out," she concludes. Worldwide, this is a new approach. She's only aware of a handful of similar experiments, in Virginia and in Germany.

In 2018 Macneale got permission from King County to seed four creeks with caddisflies, mayflies, stoneflies, and other species. It was somewhat successful. Although a lot of the introduced species didn't make it, in the different creeks, one to four species did. Among them are species thought to be sensitive to various stressors. That they survived means that conditions may be healthy enough to support other sensitive species.

It's possible that one intervention may not be enough. For now, Macneale is continuing to monitor the survivors from the 2018 reintroduction. She's eager to reseed one of the streams to see if it boosts the results, but she cautions that if you have to keep inoculating a stream with bugs, "you're not really restoring the resilience of a system. You're just patching it and needing to continue to patch it."

A Shot in the (Metaphorical) Arm

That local precedent paved the way for the other groundbreaking facet of the Thornton Creek projects. Because Hrachovec and crew had constructed the missing hyporheic zone from scratch, it was virgin territory. Inspired by Macneale's efforts to reintroduce insects, Rhodes, the microbiologist, and stream ecologist Sarah Morley, also with NOAA, inoculated the engineered hyporheic zone with microbes and invertebrates. To take the human gut analogy a step further, it was a bit like us taking probiotics—or even getting a fecal transplant—to restore gut biomes.

By phone, Rhodes and Morley explain their experiment to me. They harvested wild microbes and invertebrates in collection baskets they placed in the Cedar River Watershed, a protected area that supplies Seattle with water. It is at a higher elevation but was the closest, most similar creek they could find to what Thornton Creek was like predevelopment. "It's difficult to find lowland streams that aren't disturbed," says Morley.

They left the baskets there for six weeks to collect the hyporheic critters, then transplanted them into the urban sites. Back at the Thornton Creek confluence, I saw evidence of Morley and Rhodes's experiment: a white plastic-capped pipe, perhaps four inches wide, sticking up a foot or so from the creek bottom. They inoculated this site and the Kingfisher reach that was also restored. A few baskets they took back to the lab to document which species they had captured. They repeated the inoculations four times, once in each of the seasons.

The good news is that the sterile material in the newly created hyporheic zone was quickly recolonized by invertebrates and microbes.

Unfortunately, while the number of individuals was high, biodiversity was low. They found up to four novel invertebrate species in some stretches, but most of the diverse life forms they brought from Cedar River to Thornton didn't survive long. It was the crows-and-cockroaches phenomenon again. The species Rhodes and Morley found were surprisingly similar to those found in unrestored sections of the creek. The scientists think these dominant species likely either swept in from upstream or moved laterally underground from more distant areas of the hyporheic zone that were undisturbed by construction of the new streambed.

Morley and Rhodes aren't sure why more of their introduced species didn't make it. The science is so new that few possible explanations have been eliminated, so they are considering a wide range of hypotheses.

Microbes are simple creatures, Rhodes explains. When they pass suitable habitat, they drop out of the water current and make their home. It's a process that selects for species who are already in the stream. "They're not actively moving around like macroinvertebrates," she says. "Microbes don't really have those kinds of choices, usually. It's do or die."

Morley and Rhodes laugh. I don't. Microbiology joke?

"No free will?" says Morley.

Rhodes replies: "No free will. And very little motility."

Another possibility is that Thornton Creek was so ecologically damaged that, even "restored," it may still lack the ingredients some organisms need, or be too polluted by urban runoff. Or maybe the restored reaches are excellent but too small for newcomers to establish a strong enough foothold to outcompete the dominant gang. The restored area is "like a small island," says Macneale. "It's like how you can breed tigers in a zoo, but if you don't have the forests in which they live and hunt to release them, they're not going to make it long term."

It's also possible that Rhodes and Morley inoculated too soon after the engineering work, before vegetation needed by some critters could grow. Or maybe the donor stream's higher altitude means its denizens were never going to make it in Thornton Creek.

Interestingly, the microbes they found were much more active metabolically in the urban stream than in the healthier donor stream. That indicates they were "getting goosed to do something," says Rhodes—although they don't yet know what. It could be part of the colonization process, or they might be breaking down urban pollution, or making products they can secrete, like components to form a biofilm.

Another intriguing hypothesis is that more diverse invertebrate species survived but Rhodes and Morley's screening method wasn't sensitive enough to detect them. One way to find the tiny critters is by searching water samples for DNA that organisms shed into their surroundings through waste, sloughed cells, or gametes, material known as environmental DNA. So-called eDNA is more widely distributed through water or soil than whole bodies of invertebrates and therefore can be easier to find.

*

The water detectives had one more key monitoring question: Was the engineered hyporheic zone reducing the chemical pollution that runs off pavement, lawns, and other surfaces when it rains?

Another team of researchers used mass spectrometry to find pollutants in the stream and measured nearly 1,900. Then they introduced dye tracers at an engineered plunge pool that pushed water underground and monitored two exit points to track how long a "packet" of water stayed underground before rejoining surface flow. They followed water packets through stretches of the hyporheic seven and fifteen feet long, finding they stayed under thirty minutes to three hours or more. They sampled for the 1,900 chemicals before and after, looking to see whether the concentration of a given chemical was reduced by at least half. Just flowing downstream reduced about 17 percent of the chemicals. The short stretch of the hyporheic reduced 59 percent, and the long stretch, 78 percent. Because the time the water spent in the hyporheic was so short, the scientists think the pollutants mostly got stuck on sediments and their biofilm coatings, rather than being broken down immediately by microbes.

Hrachovec calls the chemical reduction "profound and jaw-dropping." But when considering how effective the hyporheic zone can be in reducing pollution, there are still a lot of questions. How much of a stream's water moves through the hyporheic? How long does it take for the microbes to process the contaminants stuck to the subsurface sediments and biofilms? Could a streambed be designed to capture more of the chemicals? Still, given the fact that most of these contaminants are not currently regulated—meaning there is no requirement to remove them—any improvement is a bonus. Hrachovec also points out that the study focused on "a tiny little treatment cell." He adds, "Just imagine how much good we could do if we had more of this available."

The water detectives' findings on chemical processing, as well as those on water flow and species reintroductions, are encouraging, yet they also provide a window into how complex water's relationships are with tiny creatures and its subsurface, and how difficult it is to repair what we have damaged. That's why giving a natural system as much of the space and ingredients it once had as possible will offer it the best chance of returning to a state in which it can deliver the benefits of Slow Water: flood control, water cleaning, storing water locally for dry periods.

The studies can also inform an ever-growing number of green infrastructure projects in cities worldwide. In Seattle, Lynch is now using the Thornton research to convince the city to restore hyporheic zones on four more Seattle creeks. A geomorphologist with Natural Systems Design has been building hyporheic zones into his projects outside of the city. And a restorationist working on a tributary of the Sacramento River in Northern California hopes to include a hyporheic zone.

While full biological richness may not return to an urban creek like Thornton, there are worthy degrees of improvement. Success is relative, says Morley: "Neighbors love the project. I see kids there all the time." It's a place for people, both researchers and community members, to learn about how ecosystems function.

It's also succeeded in reducing flooding. The neighborhoods adjacent to the Thornton Creek projects used to flood almost annually, closing the main road. But since the work was done in 2014, they haven't, even during large storms. By restoring space in areas that were historically floodplains, Seattle has relieved important pinch points on Thornton Creek. But it's common sense that small projects can't fully compensate for acres of lost space for water. To absorb more floodwaters and strengthen stream systems, we need to link together as many natural spaces as possible throughout human habitats. Lynch calls this concept a "string of pearls."

Nevertheless, the water detectives have already witnessed one more measure of success on Thornton Creek: in fall 2018, Chinook salmon spawned atop the new streambed and its restored hyporheic zone. "That was just really emotional," Lynch recalls. "We'd done it. We *can* restore the hyporheic zone. We *can* restore natural processes to the extent we're actually attracting salmon to the site to spawn." It was enough to lift her despair about urban "ditches." A bit verklempt, she says, "I think there really is hope for the future."

Beavers

THE ORIGINAL WATER ENGINEERS

Guerrilla gardening is the rebellious act of secretly sowing plants on blighted land to restore ecological function and beauty. In southwest England, someone has taken this concept to the next level: guerrilla beavering. Around 2008, Eurasian beavers (*Castor fiber*) popped up on the River Otter in east Devon. Aside from beavers not being the river's eponymous semiaquatic mammal (who also lives on the River Otter), their presence surprised locals because the rodents have been absent from the British Isles for at least four centuries, hunted into local extinction by desire for their soft fur, meat, and pungent glands used for medicine and perfume.

On a rainy day in early March 2020, Peter and I drive from London past Stonehenge, which stands incongruously alongside the A303 highway. We continue on far into Devon, the big toe of England that juts southwest into the Celtic Sea. Just outside of Cookworthy, a small rural town, we pull into the parking lot of a dark, rambling building that serves as the office for Devon Wildlife Trust. Wetland ecologist Mark Elliott greets us with a slightly nervous elbow tap, as COVID-19 has just been discovered in England.

Over the last several years, Elliott and other wildlife ecologists in the United Kingdom have imported beavers from mainland Europe to populate licensed studies. But those beavers were deployed in fenced-in areas. The ones in the River Otter are living wild. Elliott, who is running one of the controlled studies as project manager of Devon Wildlife Trust's beaver work, doesn't believe the Otter beavers are escapees. "I think people are deliberately releasing beavers," he tells me.

But why wield a furry, orange-toothed rodent as your stream-repair tool of choice, as people are now doing in both the United Kingdom and the United States? Beavers dig canals and build dams to create pools of deeper water so they can peruse a wider area for food and building materials with safety from land predators. Their engineering works rehabilitate stream health and function by creating areas of slow water. In the United Kingdom, people are planting beavers to help protect towns against a rise in flooding. In the western United States, people are using North American beavers (*Castor canadensis*) to help protect against drought and even wildfire. "Beavers mean higher water tables and water on the landscape throughout the dry and wet seasons," says my friend and Canadian journalist Frances Backhouse, author of the compelling 2015 book *Once They Were Hats: In Search of the Mighty Beaver.*

Other animals also played important roles in shaping the landscape for water when there were more of them and fewer of us. Bison dug or expanded wallows to mud their skin for protection from flies. The pocks they left behind acted as groundwater recharge ponds across the landscape. Prairie dog burrows provided pathways for water to move underground where the landscape was too hard to absorb rainfall efficiently. On the extremely flat surface of southern Florida, where alligators have made a remarkable comeback from near extinction, the reptile's nests elevate spots by a few feet, ultimately creating small hills called hammocks on which less water-tolerant plants can live. But along with humans, beavers are the most significant water engineers on the planet.

Backhouse draws another, less flattering parallel between humans and beavers: we're both control freaks. Yet we have very different ideas of the best way to shape water. That goes some way toward explaining why human dams sicken waterways and beaver dams restore them to health. The shortest answer is that other species evolved with beavers' interventions. Our massive engineering came about suddenly on the scale of evolutionary time. Also, beaver dams are smaller and somewhat porous; water does move through them—just at a slower rate than it would otherwise. And although water moves through human dams too, a lot of it is released on a different schedule and velocity than nature prefers. While beaver dams might seem too small to make a difference, as with other Slow Water approaches, multiple small interventions across a landscape add up to a big impact. Beaver enthusiasts—hydrologists, biologists, and ecologists who call each other "beaver believers"—are doing some detective work to measure the ecosystem benefits and water services beavers provide and to determine good places to release them.

Beaver Down

To understand how beavers' handiwork can have such powerful effects, it helps to think about what these landscapes looked like before people killed most of them, when beavers and water had an intricate, important relationship.

Prior to Europeans' arrival in North America, an estimated sixty million to four hundred million beavers lived on the continent, occupying nearly every headwaters stream. As nature's original Slow Water engineers, beavers shaped North America with their dams, creating a much wetter continent than the one we know today, spreading and slowing water wherever they went. Their ponds covered more than three hundred thousand square miles, turning one-tenth of the continent into rich, ecologically diverse wetlands that supported numerous other species—textbook behavior of ecological engineers. The longest documented beaver dam, found via satellite in northern Alberta, measures 2,790 feet—twice the length of Hoover Dam.

Many Indigenous peoples across North America hunted beavers too, but they typically regulated their harvests with beliefs of interdependence among species and recognition of the spirit in nonhuman beings. Together these values constrained their take, according to *Once They Were Hats*. Some groups even viewed beaver as kin, or part of their clan. Then, "five hundred years ago, the nature of this relationship changed dramatically, as foreign economic incentives and technology rewrote the rules of engagement."

After European fur traders arrived, beaver numbers plunged to less than 1 percent of their historical population by 1900. This was no accident. In the mid-1800s, Hudson's Bay Company ordered its trappers to hunt out all beavers in the Northwest, creating a "fur desert" that would elbow out competing companies. Beaver numbers across North America dropped to as few as a hundred thousand, according to Ben Goldfarb, author of the charming book *Eager: The Surprising, Secret Life of Beavers and Why They Matter* (2019). The animal's decimation fundamentally altered the plumbing of much of North America. Without beavers and their dams, water moved to the ocean more quickly, scouring streambeds, deepening their channels, and reducing slow phases and spaces where surface water could recharge aquifers. Groundwater levels fell. The North American landscape became a drier, less ecologically diverse place.

Across the big pond, in the United Kingdom, people had lived alongside giant beavers since the Stone Age. Later humans shared space with the modern Eurasian beaver. Both humans and beavers retreated to the south with the advent of various ice ages, and both returned after the Last Glacial Maximum, about 9500 BCE. As the climate warmed, what was once tundra became mixed forest, and later "woodland," a partially open blend of trees and shrubs. During this period, wetland plants and animals adapted to rely on the habitat beavers created, Elliott tells me on my visit to Devon. Consider willows, he says, which coppice (resprout) in response to beavers chewing them: "You get a change in the canopy structure because the beavers are cutting down the bigger trees, and you're getting this lovely fresh regrowth of younger stuff. As the beavers are moving around, the vegetation is changing in response to what they're doing."

But people were also interested in woody plants and began clearing forests to farm as far back as the Neolithic period, about 4300 BCE. Fast-forward to the start of the Industrial Age, when logging ran rampant to build ships and fuel steam engines. Around the same time, Britons hunted the last of the native beavers.

The lowlands were already prone to flooding during big winter storms. Removing trees and beavers exacerbated the problem, and in recent years, climate change and development have made it worse still. UK Meteorological Office records stretching back to 1862 show that six of the ten wettest years have occurred since 1998. In just the last few years, the UK has seen damaging floods in Cumbria, Yorkshire's Calder Valley, South Wales, and Shropshire. Even under a low-impact climate change scenario, flooding is predicted to increase in 85 percent of UK cities with a river, including London. Under a high-impact scenario, some cities and towns could see the quantity of water per flood double.

After beavers were exterminated, or nearly so, in the United Kingdom and North America, humans exacerbated damage to streams and rivers, as in Seattle, by speeding water off the land even faster. They straightened waterways' meanders and even moved them by digging new channels to better conform with property lines. Add to this the hard grazing by cows and sheep that rips out grass from the roots, plus the animals' tendency to stand next to waterways for hours, tamping down the soil, and you end up with thoroughly ill streams.

The absence of beavers from the landscape over several generations of humans means that people today are accustomed to a landscape devoid of beavers—and in fact larger animals generally. It's the "shifting baselines" phenomenon again. The idea of coexisting with other animals, relinquishing space to them, weighing their needs as equal to our own, is largely anathema to the dominant culture in which human supremacy goes largely unquestioned.

We also seem to be uncomfortable with change. Beavers' handiwork restores a more natural landscape, but many people think their tonic looks messy: trees felled across creeks, piles of sticks forming dams that look a lot like slash piles from clearcut logging. But the neat, straight channels we've built are ailing systems. Beaver-built "roughness," as ecologists call it, makes possible Slow Water and all its benefits. The logs and detritus that Mike Hravochec and Natural Systems Design placed carefully across Thornton Creek in Seattle are an imitation of beavers' work—a mimicry that Miwok people in California made earlier to slow water on steep slopes, infiltrate it underground, and create wet microhabitats.

The same pursuit of Slow Water that inspired the Thornton Creek restoration crew is also luring beaver believers across the US West. They are hoping to reap these benefits for local water storage, a goal more important than ever as snowpack and glaciers are, for the most part, not capturing the quantity of winter water they did in years past. Beaver dams slow water, and sediment drops out of the flow and begins to raise the stream bottom. Eventually, water can access the floodplain again and disperse across the landscape, slowing it further, reducing scouring, erosion, and flash floods. Then microbes have time to break down pollution and water can infiltrate underground for storage.

In the United Kingdom, people are looking to beavers to protect against increasing big floods. Beaver dams slow water rushing down streams from heavy rains. Scientists call this "attenuating the peak flow." It's similar to the principle of flattening the curve during the coronavirus pandemic. By quarantining, societies slowed the spread of the virus, reducing the peak numbers of people hospitalized, ideally to a level healthcare workers could handle. Because beaver dams are somewhat porous, much of the water detained will ultimately flow down the river. Some evaporates from the pond, some percolates down into the groundwater system and may rejoin the river later, further downstream. The net result is that water travels downstream over a longer period, reducing the possibility of a huge wave that overflows banks and floods towns.

Back from the Past

In Cookworthy, Mark Elliott loans us Wellingtons and we caravan out to his study site on nearby private land, owned by retired farmers John and Elaine Morgan. The beavers' domain is a fenced-in, 7.4-acre plot on a branch of the Tamar River headwaters. From the road we walk a short distance into a field. Elliott deactivates the electric fence surrounding the site, and we step over a strip of wire mesh that deters the beavers from digging to freedom.

This site, called West Devon, centers on a small stream, across which the beavers have felled numerous trees. Willows alongside have little angled bite marks and are furiously regenerating new stalks in response to the munching. A metal weir across the stream with a V-shaped notch measures the speed of flow entering the site, and another measures it upon leaving. Together the two weirs track the rate at which beavers slow the flow and reduce flood peak in the area. The ground surrounding the stream is saturated, and the deep muck sucks at my loaner wellies. Water is falling from the sky, persistently, but Elliott doesn't raise his hood. About forty-five minutes in, he remarks, "Nice it's not raining today!" If you study beavers, especially in the UK, maybe you just get inured to being wet.

In 2011, Elliott and his colleagues kicked off this site by introducing a couple descended from Bavarian beavers. This project was authorized by Natural England, a conservation arm of the national government. There are now multiple fenced trials across the country, including those in North Yorkshire, Cornwall, Essex, Somerset, Gloucestershire, Cumbria, Norfolk, West Sussex, and Dorset. Here on the West Devon site, beavers now number five, and they have constructed thirteen dams, creating thirteen ponds across their territory between 2011 and 2016. In slowing the water and spreading it laterally from the original stream, they have expanded the wetland area from about one thousand square feet to twenty thousand square feet. Together the ponds hold up to 264,000 gallons of additional water above ground. Since then, the beavers have not built additional ponds, but they continue to adjust the height and width of their dams, and the volume of water on the site continues to rise.

Anecdotally, the landowner, John Morgan, has told the researchers that the road immediately downstream of the site used to flood regularly during heavy rain but hasn't done so since the beavers were introduced. The water stored on site also helps sustain the stream during dry spells. Summer 2016 brought drought, and stream flow coming into the complex completely dried up. Yet water continued to flow out, keeping the stream below viable.

Elliott's first love is herpetology, and like a proud parent, he gleefully points out dollops of frog spawn while cautioning us not to step in them. They look like giant loogies. The beaver-created wetland has led to a huge increase in number of different species and individuals, and the frogspawn is a good indicator. When the project began in 2011, Elliott counted just ten clumps of frogspawn; in 2016 he counted 580. In turn, this robust food supply attracts critters such as grass snakes, herons, and kingfishers. Other newly returned and increased species include bats, willow tits, and aquatic insects and other invertebrates. Plant communities have changed too. As the beavers have cut willows, the area has become more open, allowing liverworts and grassland plants to come back, says Elliott. He's been particularly excited to see species such as purple moor grass recolonizing, which is characteristic of a rare grassland habitat known locally as Culm.

*

After touring us through the West Devon beavers' greatest hits, Elliott leaves to pick up his kids. Peter and I head southeast toward the English Channel to visit the guerilla beavers' domain. We wend our way down small lanes just big enough for a single car. These once-dirt carriageways have been tamped down over the centuries, sinking below the fields they bifurcate. Hedgerows that line the fields tunnel us in further, adding to a cramped feeling that turns to adrenaline when we nip around a tight corner and are suddenly confronted with a fast-moving local car. Everything seems tightly controlled here, without a lot of room to spare.

Eventually we cross the River Otter and arrive at Otterton Mill, where a water wheel has been churning for a thousand years. Today the wheel still grinds flour and serves as the centerpiece for a café–restaurant–arts center. The mill sits at one end of Otterton, a tiny old English village with just a few streets. Here we meet one of Elliott's colleagues, Jake Chant, field officer of the River Otter Beaver Trial. A sinewy guy with a wide smile and a fast patter, he's perpetually available for beaver emergencies. His out-of-the-office automated email response offers this: "If you need urgent support with beaver management please call me on . . ."

Pulling out a chair at the café and sitting down, he echoes my impression from the drive, saying, "Nature really gets no space here." It's hard to imagine a landscape more coiffed by human hands than the British countryside. "Even our national parks are not wilderness areas. They're just a bunch of farmland and people have pressed pause on development."

Into that ethos, enter the unplanned beavers. When their presence was captured on video, proving their existence, the UK government plotted to remove them, arguing that they were a "nonnative animal." It was that attitude that had driven folks to guerrilla beavering in the first place. Local people protested, petitioning for the beavers to stay. Devon Wildlife Trust joined the cause, helping to convince administrators to make these wild beavers part of an official trial licensed by Natural England. They got conditional approval, in partnership with University of Exeter researchers, for a five-year study that ended in 2020 and covered almost one hundred square miles of the River Otter catchment. In 2015, the river held two beaver families comprised of nine individuals. Now there are eighteen family groups on the river, says Chant, supplemented by a couple of authorized imports to increase their genetic diversity. Chant has tracked beaver activity from the estuary to the headwaters, a forty-mile stretch.

He says the beavers explore streams throughout the watershed, and good-quality food, such as their favorite willows, seems to factor heavily in their choice of where to settle. In the first few years, the beavers were in the lower reaches of the river, where deep water meant that they weren't inspired to build dams; they dug their homes into the side of the riverbank. More recently, as their population has increased and young beavers have set off to find their own homes, they have moved into shallower areas and have begun to build dams in headwater streams, where their work is more likely to protect downstream humans from floods. About half the beaver families have built dams, mostly in an upstream tributary.

The River Otter Beaver Trial study, published by the University of Exeter, found the beavers' expanding numbers and engineering activities have had positive benefits for wildlife, fish, ecotourism, and cleaning pollution from water. And the thirteen dams at Elliot's West Devon site and the six dams on the River Otter upstream of the village of East Budleigh significantly reduced peak flood flows. During storms, peak flows were 30 percent lower, on average, leaving the sites than they were without beaver dams. These effects were true even in saturated, midwinter conditions. Follow-up research confirmed similar results across multiple beaver sites in England. Even during large storms, beaver dams still reduced flooding. In fact, at the West Devon site during Storm Frank in December 2015, dams had even more impact: outflow was three times lower than inflow.

In 2020, the government decided the River Otter's guerilla beavers could stay, concluding, "The five-year trial has brought a wealth of benefits to the local area and ecology, including enhancing the environment at a local wildlife site, creating wetland habitat, and reducing flood risk for housing downstream. They will now be allowed to remain there permanently and continue to expand their range naturally, finding new areas to settle as they need."

Devon Wildlife Trust was elated, saying in a statement, "This was a landmark decision and one of the most important moments in England's conservation history. It was the first legally sanctioned reintroduction of an extinct native mammal to England." Meanwhile, guerrilla beavering continues. Beyond the River Otter, other populations of free-living beavers were spotted on rivers in Kent, North Somerset, and Wales. As of the time of this writing, Natural England is planning a strategy for managing wild beavers and potential further releases into the wild across England.

Today, beaver reintroductions are increasing in popularity around Europe. UK restorationists were inspired by work in Bavaria, Germany. Scotland has also been reintroducing beavers, officially welcoming them back in 2016 after several centuries' absence and awarding them legal protection in 2019. There, the animals are living wild in Knapdale Forest in Argyll and along the River Tay. Eurasian beavers, including those in the UK, are now estimated to number 1.2 million across the continent. Beaver believers in the western United States have also been an inspiration to the Brits, even though the Americans are primarily interested in enlisting the critters to fix the opposite problem: drought.

Beaver Dreams in Washington

"We've got one!" yells David Bailey, assistant wildlife biologist for the Tulalip Tribes, just after 7:30 a.m. on a clear Wednesday morning in September. He's smiling, dimples peeking through his thin beard, as he returns from scouting a small pond at the end of a conifer-bedecked suburban street in north Seattle. His colleague Molly Alves and I pull on rubber boots from the back of their big, dirty Suburban and walk across muddy ground to the edge of the pool abutting Stickney Lake. The long angle of morning light is warm and soft, highlighting yellow fall leaves and cattails that frame the pond. Lily pads dot the surface. Bailey hails from Eastern Washington, while Alves came from farther afield, moving here from Vermont a few years ago to take this job as wildlife biologist and resource manager for the Tulalip Tribes. Together they bend over and grab handles on a large cage, straining to pull the heavy animal out of the water.

The beaver is in a metal wire pouch with framed crossbeams for lifting and maneuvering. Alves and Bailey carry the animal between them, breaking heavily through the brush with their load. The beaver is moving around a little bit, clearly concerned but not frantic. When they reach me, they stop for a rest. "You've been eating! Very well," Alves says, alluding to the beaver's portly frame. Brushing her long, straight hair back over her shoulders, she appraises him with a practiced eye, estimating that he may weigh fifty pounds. The beaver scoots around, looks briefly toward me, then positions himself in profile.

He sits on his tush, flat tail out in front of him, webbed back feet with long claws resting on top, lush fur sticking out through the cage. He holds his little forepaws together in front of his chest, as if in prayer. His nose appears larger than his front hands, and the eye that I can see is quite small, dark brown, and squinted half closed. He may be tired from a night spent trying to get out of the trap, Alves theorizes. She says I can touch him, and I reach out a finger to his thick, wide tail. It feels a bit like snakeskin, finely textured with a hatch pattern, and alive: firm, with a tiny bit of give. His beautiful coat is shiny with undulating shades of black and brown, impossibly soft, yet composed of hairs coarser than, say, a rabbit's or a cat's. Although I am against wearing fur, it's easy to understand the allure of this animal's amazing coat and how it drove the fashion frenzy that led to its near demise.

Unlike in the United Kingdom, where beavers are so scarce that restorationists must import them from Europe, in North America the animals managed to hold on after the human assault. Today their numbers are growing. Although no federal agency systematically surveys the animal, author Goldfarb says biologists estimate there are about fifteen million beavers in North America today. Unfortunately, as they return to their homelands, they often run into trouble. During their centuries of near absence, another animal moved into their preferred habitat: us. Where a truce between the two species remains elusive, biologists catch the beavers, as Alves and Bailey are doing this day, and relocate them into more remote terrain where their skills can be put to good use.

In another installment of "Beavers: They're Just like Us!," beavers and people tend to want to live in the same spots—flat land near water. When European trappers arrived here, the highest concentrations of beavers were in low-lying, wide valleys. After trappers quickly killed the animals, the big wetland complexes dried out. The flat land, rich with silt from millennia of water inundation, was very attractive to farmers. As they moved in, settlers exacerbated the drying-out by draining remaining wetlands, diverting water from rivers to irrigate crops, introducing cattle and sheep that grazed grass to the nub, and killing predators that would otherwise keep grazing animals moving along. As time passed, some of these farms and ranches gave way to strip malls, housing developments, and airports.

Like us, beavers live in family groups and work together to supply themselves food and shelter. Juveniles live with their parents for a couple of years, working on dams and lodges, felling trees for food and building materials. But at two years old, it's time for the youths to set out, find mates, and start their families elsewhere. In these efforts, they sometimes come into conflict with people, who resent it when the beavers flood a field or house or fell a favorite tree.

California transplants beaver into El Dorado National Forest via parachute in 1950 (Source: CA Dept. of Fish and Wildlife).[16]

FIGURE 4.1 California's Busy Beavers poster.
© California Department of Fish and Wildlife

*

Although many beavers have paid with their lives for crossing human desires, wildlife managers in the western United States began
relocating "problem beavers" as far back as the 1920s to take advantage of their skills repairing watersheds. In a 1950 paper in the *Journal*

of Wildlife Management, Elmo W. Heter with the Idaho Department of Fish and Game wrote that beavers "do much toward improving the habitats of game, fish, and waterfowl and perform important service in watershed conservation."

However, beaver redeployment was somewhat limited in Idaho at that time by a lack of roads into the backcountry. The Idaho Department of Fish and Game came up with a solution inspired by World War II airborne divisions: parachuting beavers into the wilderness. Wildlife managers built a crate designed to deconstruct upon landing. After experiments with dummy weights, they enlisted as their test subject an older male beaver, "whom we fondly named Geronimo," writes Heter. They dropped him again and again to make sure the system worked. "Each time he scrambled out of the box, someone was on hand to pick him up. Poor fellow! He finally became resigned, and as soon as we approached him, would crawl back into his box ready to go aloft again." (Or maybe he enjoyed the view from the sky, or the thrill of the jump?)

The researchers decided that Geronimo deserved just reward for his service: "You may be sure that Geronimo had a priority reservation on the first ship into the hinterland, and that three young females went with him," writes Heter. "Even there he stayed in the box for a long time after his harem was busy inspecting the new surroundings. However, his colony was later reported as very well established." In fall 1948, the researchers dropped seventy-six beavers in the Idaho mountains, suffering just one casualty, a beaver who managed to get out of the crate while it was still in the air.

The California Division of Fish and Game (later renamed California Department of Fish and Wildlife) had also been deploying beavers as watershed engineers. Between 1923 and 1950, staff distributed 1,221 beavers into watersheds across the state. California Fish and Game biologist Donald Tappe explained why in a 1942 report: "It is now understood that soil erosion and shortage of water in some places resulted from the destruction of the beavers which formerly built, and kept in repair, dams on the upper reaches of many streams."

California wildlife staff, inspired by their Idaho colleagues, deployed the last of those beavers by parachute to El Dorado National Forest, between Sacramento and Lake Tahoe. A poster from 1950 shows a cute line drawing of a beaver (sans crate) parachuting into a wilderness scene with mountains, stream, deer, and other beavers building a dam. The caption explains that "California's busy beavers" are being relocated from farm areas to mountains to "save water for fish, wildlife, and agriculture."

Beaver Jedi Mind Tricks

Despite this history, many people's first instinct when facing a "problem" beaver today is still to kill it. A US federal agency euphemistically called Wildlife Services claims its mission is to "resolve wildlife conflicts to allow people and wildlife to coexist." In practice, though, wildlife often ceases to exist. The agency reported killing nearly twenty-six thousand beavers in forty-three states during 2020. Private citizens are also allowed to kill beavers on their property.

Yet killing them is only a temporary fix in most cases. "Removing beavers just opens up prime beaver habitat for the next family that comes across it," Alves says. "We always tell landowners that trapping is a short-term solution to a long-term problem. If you have beavers now, you're going to have beavers again."

Because of this persistence, beaver believers such as Alves first try to create détente between people and beavers in place. Beavers and humans are both territorial. Yet if people and beavers can come to an understanding over water management, they can live peacefully as neighbors for years, until the beavers run out of food or their pond silts up. One of the key tools for lowering water to avoid flooding things people care about is called a pond leveler, a.k.a. a "beaver deceiver." It was invented by a wildlife manager famous in beaver circles for this device, Vermont-based beaver herder Skip Lisle.

To see this tool in action, I visit Big Spring Creek on the outskirts of King County in Washington, near the town of Enumclaw. Showing me the site are two beaver wranglers from Beavers Northwest, an NGO that negotiates peace treaties between beavers and people. They are Ben Dittbrenner, cofounder of the organization and an ecologist who initiated relocations with the Tulalip Tribes, and Elyssa Kerr, a restorationist and environmental educator who's now the organization's executive director. Their work on Big Spring Creek is part of a project with King County and the Army Corps of Engineers to restore habitat for threatened Coho salmon.

I climb into the back of Dittbrenner's gray Passat, next to a child's car seat. A rooftop cargo box is loaded with tools of the trade: metal poles with instruments to measure water flow and potato rakes, a sort of hoe with tines. Driving south, we are soon outside of Seattle, making our way down green highways and through small towns. We pull onto a dirt road and park alongside a small farm. Dittbrenner, with a beard just turning gray, opens the hatchback so we can don the requisite gear. I pull on the chest waders Kerr brought for me. This sexy item, de rigueur for wetlands ecologists, is a pair of waterproof pants that rise nearly to the armpits, held in place with suspenders. Kerr somehow manages to look graceful in her Cabela's chest waders, blond hair braided to the side. Out of the front pocket she extracts her phone, clad in a plastic sleeve for safe note-taking over water. Chest waders have integrated pajama feet made of waterproof neoprene. Over the footies go tough, construction-style boots—also waterproof. As Dittbrenner arms me with a potato rake, completing the picture, I feel like a reenactment of Grant Wood's *American Gothic*. We start the half-mile walk down the road to the stream, through invasive canary grass taller than our heads. I soon discover an important use for the potato rake as a kind of sounder, testing in front of me for depth, which can drop off suddenly into invisible crevasses.

Originally this area was a giant beaver complex, Dittbrenner tells me. When farmers moved in, they drained the remaining wetlands and moved the stream uphill to the edge of a farm field, which caused regular flooding. That's part of why this area was selected for salmon habitat restoration—to solve that problem too. The restoration has made extra space for slow water, which has reduced flooding overall. But the beavers' work also ended up flooding the edge of an adjacent landowner's field. That's when the government agencies called Beavers Northwest to help keep both beavers and people happy.

The day is perfect, with a whisper of remnant summer in the deep-blue sunny sky of early fall. As we get close to the creek, walking becomes more difficult. Just six years ago, this land was still an open farm field cut by a straight, channelized drainage ditch, banks denuded. The restorers reintroduced curves into the stream and created backwater ponds for slow water. They planted thousands of willows along the banks and alders behind, and dropped logs across the channel to simulate a natural system where trees fall across the stream. To the beavers, it suddenly looked quite homey.

"Beavers just moved in immediately," says Dittbrenner.

Already, willows are so thick along the banks that we have to crouch to move through them. Around each gnawed branch, several small new branches fan out around the bite, just like the ones Mark Elliott showed me in Devon. Spiky pines, wild native roses, and blackberries snag our clothes and poke our skin. When the going becomes impassible, we step into the stream, water tea-colored with tannins. I slog along carefully, wielding the potato rake in front of me to serve as my eyes, stepping up onto invisible underwater logs, then down into deep holes with mucky bottoms. It's clearly not an environment designed for humans. Traversing it is both physically and mentally exhausting.

At various points along the creek, Kerr and her crew have installed the pond levelers: black corrugated plastic pipes lying just under the water. They prevent flooding by allowing water to pass through dams or other beaver-created obstacles, partially draining the pond, reducing its water level and, therefore, nearby flooding. The tricky part is if beavers hear the sound of running water, they'll plug the hole. To avoid this, the outlet of the pipe needs to be underwater, where it won't make a sound. That can be challenging because water levels naturally fluctuate with the seasons. In some cases, beaver believers build an impenetrable cage around the outlet, and frustrated beavers eventually come to accept that they won't be able to stop the sound. This site has a couple of cylindrical cages around the ends of the pipes.

Working with these animals never becomes rote, though. Beavers are individuals, all with different temperaments and proclivities. Like people, they are problem solvers and highly adaptable to different situations. "Once they learn how to overcome something like a human intervention, they'll replicate it every single time, and they teach it to their young," Dittbrenner says with clear admiration.

Beavers continually surprise the people who work with them. Although their classic needs are trees and water, they've been found living in grasslands or tundra, and even in a dry ditch. In open agricultural areas without a lot of vegetation, Dittbrenner has seen them build dams with canary grass they dig up from the root wads. To pack it, they mine the mud from just above the dam "because that's the slowest water and the mud is settling, so it's this replenishing mud supply." He explains their technique: "They swim down and grab [some mud] and hold it on their chest as they swim up, and then they walk up on their feet with their tail like a little support so they don't fall backwards. And then they'll pack it into the dam."

Dittbrenner and Kerr show me one of the larger dams beavers built on this stream. On the surface of the pond they've created, sunlight dances on filamentous swirls of pollen. When working with their preferred media, wood, beavers are expert crafters, says Dittbrenner: "When you pull a dam apart, it's just amazingly interwoven." They don't just throw a bunch of sticks together. "They're sliding them in, and then they do another one, and they slide it in, and as it gets pushed down, it just kind of naturally weaves itself."

Later, walking back to the car, we cross a small section of open meadow. Although we choose a path through the grass about thirty feet from the creek, the land is boggy. The beavers have created this high water table, the groundwater pushing toward the surface. It's a habitat zone that humans have eradicated in many places. More than a third of endangered or threatened species in the United States rely on wetlands. And riverside corridors of vegetation are particularly valuable for species migrating to adapt to climate change. In this clearing, redwing blackbirds dip by, their red squares flashing. We spot a tiny nest woven from grasses anchored to a few of the reeds. Cattail "corndogs" are breaking down, releasing pale fluff. A bright-green Pacific tree frog clasps hands and feet around a branch speckled with goldenrod and gray-green lichen. The frog's black markings

slope downward from the corners of her amber eyes, blinking back
at us.

Fortunately, when beavers recreate this habitat, there are enough
wetlands remaining that the areas tend to be quickly recolonized by
their native species, says Dittbrenner. "There's a pretty rapid inocu-
lation of life from wetland to wetland." Alves has seen that phenomenon too. Her crew with the Tulalip Tribes has set game cameras near
potential beaver reintroduction sites and tracked just a few squirrels
in a whole year, she tells me. After reintroducing beavers, "two years
down the road, we have bobcat and bear and cougar and otters and
salmon returning to some of these sites."

Keeping human neighbors at Big Spring Creek happy, though, requires ongoing tinkering. And some of it is driven by fear of change,
not by actual problems, Kerr explains. A lot of people "see beavers
moving into an area and they panic." This response raises questions
for Dittbrenner. "What's our long-term management strategy, right?
Are we going to put pipes in every single stream?" Tolerance and
adaptation work better when they go both ways. Neighbors at Big
Spring Creek will accept only a six-inch rise in water, which requires
really tight control, he laments. If they could live with even a foot,
the managers would have much more flexibility. Small things—like
building a tiny berm to direct water away from a vulnerable field, or
allowing a bit of water on the lawn occasionally, rather than being
able to mow it at any time—can give beavers and natural systems
much more leeway to work their alchemy.

Moving Day

Where people just won't abide beavers at all, wildlife managers
such as Alves and Bailey relocate them, which is their plan for the
suburban beaver I met. But human intolerance toward beavers' existence is not necessarily limited to location. Some people object to
beaver relocations too because they fear the dams will block remnant salmon runs. That's ridiculous, according to Alves. Beavers and
salmon evolved together. An increasing body of research shows that
beavers, in their role as ecosystem engineers, help salmon popula-

tions to flourish. They create ponds of clean, cool water and expand gravel areas for spawning. Young fish grow faster and reach the sea healthier if they spend time in beaver ponds.

In fact, the fish also help create the shape of a healthy stream. To make nests for her eggs, a female salmon digs a depression by turning on her side and whipping her tail up and down, stirring up gravel from the river bottom so the current carries it downstream. This motion creates a pit for eggs in the hyporheic zone, about one foot deep. Just downstream, the excavated material forms a little hill. A male fish swoops in to fertilize the eggs, and the female moves a bit upstream to do it again, her flapping covering eggs in the first pit while digging another. Across the streambed, the nests look like a series of moguls on a ski slope. Aquatic ecologists modeled the impact of these digs over millions of years and concluded that spawning may have lowered some riverbeds 30 percent more than without salmon digging, significantly shaping the channels and water flow.

Still, human fears for salmon—which are a bit ironic, given that we have endangered them via our water engineering and land-based development—continue to bedevil human–beaver relations. Alves and Bailey had to catch that big male because the state won't allow pond levelers to reduce flooding in the area near Lake Stickney due to concerns about salmon, arguing that fish can't pass beaver dams at low flows. That view ignores the science, says Alves. It's true they can't pass at low flows, she acknowledges, but "that's why salmon run when it starts raining."

Human-built infrastructure is the real culprit here, she asserts. People have confined streams into culverts, dramatically decreasing the space salmon have to move. Then, "when a beaver builds a dam in a culvert—this tiny tube that you've put a stream in—well yeah, it's going to prevent salmon from migrating upstream." Alves adds, "We always put air quotes around 'nuisance beavers' because they are not the nuisance. It's the infrastructure that is enabling them to get themselves into trouble with human values."

In 2012, the Washington state legislature passed a law allowing beavers to be released east of the Cascade Range summits but not to the west. For that reason, some early beaver relocations in western

Washington were on tribal and federal lands. Tribes are sovereign nations, so they can decide if they want to release beavers on their land. The Tulalip Tribes was among the first to try it, motivated by a desire to restore their salmon runs. Dittbrenner used to work with them and developed the beaver relocation project now helmed by Alves. Since 2017, state law has been amended to allow relocations in the western part of the state also. Now Alves's crew has permission from the US Forest Service to release beavers on its land within the Snohomish watershed. They are hoping to expand onto state land too. Within these areas, they look for places where beavers have the ingredients they need and where their labors can repair stream hydrology.

With the big beaver sitting quietly in the back of the Suburban, cage covered by a sheet to reduce stimulation, we drive to the twenty-two-thousand-acre Tulalip Tribes reservation, where more than half its 4,900 members reside. It has a series of residential neighborhoods and an elegant, large administrative building. It also has a salmon hatchery, which is our destination today. At this time of year, the salmon have been moved out of the long concrete raceways. This will be our beaver's temporary home. The team has built little platforms out of the water for the beavers to sleep on and provided some nesting materials. The water in the raceway is festooned with boughs and branches for snacks and simulated habitat. Because beavers are family animals, the team holds them here until they can capture the whole clan for relocation. Single beavers are housed with a member of the opposite sex to give them an opportunity to pair up before heading out to their new home. The human matchmakers don't always succeed. Some individuals don't hit it off, so the people have to try again with other beavers. They'll figure out the fate for today's fellow after an examination.

First, Alves and Bailey tag the big beaver. Alves steps on one side of the pouch cage to maneuver his head into the opposite corner, where they can pierce his ears with little colored streamers. This guy is the forty-second beaver they've caught in 2019, so he is No. 19–042. The fluorescent orange tag they shoot into his left ear represents 4, and the blue in the right means 2. In this way, they will be able to identify him in the field later without recapturing him. They tweeze a small tuft of hair for ongoing genetic research projects.

Then it's time to determine the beaver's sex. Although Alves, Bailey, and I have been using male pronouns for the beaver, it's impossible to tell by looking whether a beaver is female or male. To find out, you have to get up close and personal. To conduct this delicate procedure, they usher the beaver from the pouch cage into a "beaver bag," a conical canvas tube with a hole at the tip like a pastry piping bag. Except its opening doesn't dispense frosting but allows the beaver to breathe. They wrap it tight so the critter can't struggle much and then invite me to sit down on a milk crate and hold the big, heavy, warm bundle while they check the sex, advising me to keep my hands away from the teeth zone.

Alves coaches Bailey on where to apply pressure to push the anal glands out from the cloaca, an orifice for both eliminating waste and reproduction. Bailey starts squeezing, and a stream of pale liquid arcs out. Alves says this is urine, not the target goo that identifies sex. A lot of urine. I inch my arm back from the business end, hoping to avoid testing my coat's waterproofing while trying not to lose my grip on the heavy animal or jiggle Bailey's fingers away from his goal. Finally, he gets the gland to emerge and squeezes in a milking motion to release oil from the castor sac, a valuable commodity in the past, used as medicine and as a base in perfume. Researchers determine gender by color, viscosity, and smell. Old cheese signifies a female beaver, motor oil a male.

"Ah, the glamour of my job," laughs Alves, leaning in to absorb a bit of the oil with a paper towel. Determining the sex of a beaver is a strange talent. "It's definitely weird and kind of makes you blush when you tell people what you have to do every day," she says. We each give it a cautious sniff: it's not awful, but it's getting there. The pungent odor does bear some resemblance to motor oil. This is likely the dad of the clan they'd caught at the same lake earlier that week, a mom and two "teenagers." He will be reunited with his family at the hatchery for a few days until Alves can line up a relocation spot for them.

Alves is now sharing her strange talent and other beaver knowledge elsewhere in the western United States. In California, beaver believers are fighting the misconception that beavers were never native to the state because they'd already been hunted out when an influential guide to indigenous critters was written. But tribes in California, like those in Washington State before them, are leading. The Tule River Tribe, whose land is south of Sequoia National Park in the central part of the state, has contacted the Tulalip Tribes, Alves tells me, seeking its advice on beaver restorations. The Tule River Tribe has received a grant from the US Fish and Wildlife Service to begin its pilot beaver relocation project. For the next year or so, members will assess potential spots on the reservation; beginning in 2022, they will relocate a handful of beavers to those areas. About fifteen other tribes in California have "expressed serious interest" in relocating beavers, says Alves. The Tule River Tribe pilot study will define how to do that and "hopefully set the precedent for these other tribes. Very exciting stuff!"

Drought-Busting Power

One of the main goals for beaver reintroductions in the western United States is to heal sickened streams, thereby increasing the amount of slow water on the land and underground. The science demonstrating the benefits of beavers' hydrological work is young but an active field of research. Researchers in 2015 found that the average beaver pond contains 1.1 million gallons of water and stores another 6.7 million gallons of water underground. Beaver water complexes can even act as a fire break against the megafires that climate chaos is wreaking here. And given that desiccated plants are volatile tinder for fires, it's possible that a widespread return of beavers could help reduce fires both by keeping plants better watered and by providing more local evaporation from the ponds and transpiration from plants to fuel local rain.

Slowing water on the land and underground is particularly important in a climate-warmed world with less snow, which is already a problem in Washington. Under the business-as-usual climate scenario, the Cascades are projected to lose nearly 100 percent of the snowpack by 2080. "That's disturbing because that snowpack is primarily what keeps a lot of our mountain streams flowing through the summertime," says Dittbrenner. For his doctorate in forest and environmental science, he studied how beaver reintroductions can help make up for a climate change–altered water supply. He relocated seventy-one beavers over two and a half years and tracked changes in water storage at their new homes.

On average, the beavers Dittbrenner studied stored 2.4 times more water underground than their ponds held on the surface. The amount of water stored at a particular site varied due to elevation, topography, rock and soil types, precipitation patterns, and water vapor released by soil and plants.

Another significant finding was the difference in the beavers' impact between water basins currently dominated by snow and those dominated by rain. In rain basins, beavers' water storage will increase summer water availability as the climate changes by up to 20 percent, compared with up to 5 percent in snow-dominated basins, Dittbrenner found. The difference between the two is largely because rain-dominated basins tend to be in lowland areas that are wider and flatter, meaning that compared with steep mountain streams where snow falls currently, beavers have more space to store larger quantities of water. Plus, the minimal slope reduces water speed, resulting in longer holding times and increased groundwater infiltration.

Although the beavers' 5 percent contribution in snow basins sounds small, it's meaningful: more than four billion gallons in one basin. That quantity of water storage in just that one basin is almost one-quarter the capacity of the Tolt Reservoir that serves Seattle, Dittbrenner estimated. In areas further inland with different precipitation patterns and increased space for habitat, beavers could potentially store much more water to replace what was once snow, slowing water on the land and delivering it to people during the dry season.

*

Researchers in Northern California have also been studying beavers' contribution to groundwater recharge. I lucked into seeing scientists at work collecting data in Childs Meadow, an alpine landscape near Lassen Volcanic National Park and the tiny town of Mineral, where my great-grandfather built a cabin in the 1930s. I visit nearly every summer and know this area well. In June 2019 my family vacation happened to overlap with the researchers' field season.

It turns out, the alpine meadows I've long loved here are not well. For more than a century, they've been grazed hard by cattle, and the streams were straightened, leaving the land much drier than it was historically. Although I'd always thought of mountain meadows as dry grasslands, in fact, they are wetlands, says Sarah Yarnell, research hydrologist for the Center for Watershed Sciences at UC Davis. When beavers and willows were removed and grazing cattle introduced, mountain meadows like this one changed from seasonally flooded land with braided channels into single channels cut into the earth, with wide-ranging consequences—not just for water, but also for greenhouse gas emissions. One study from the Rocky Mountains estimated that beaver meadows once stored about 23 percent of the carbon in this type of landscape. When people hunted out beavers, converting wet meadows to dry grasslands, carbon storage decreased to today's average of about 8 percent.

Childs Meadow is around 550 acres; it takes fifteen minutes to drive its full length on the paved highway alongside. When I see tiny people in the distance, I park and walk toward them, passing through bands of grasses and gum plants—pale green, reddish, dark brown, bright green—the species dictated by slight alterations in elevation and access to the water table. Along the horizon, Jeffrey, sugar, and Ponderosa pines ring the meadow. Just beyond them peep the tops of snow-covered peaks. It's June and the alpine flowers are busting out: little golden buttercups, fuzzy pink pussy paws, and at the base of trees, waxy red snow plants pushing through pine-needle duff. I watch my step to avoid cow patties and delight in periwinkle-blue butterflies flitting about.

The researchers are comparing four sections of land used in different ways, rating their carbon storage, water-holding capacity, and habitat for sensitive species. In the first section, cattle still graze free in one part of the meadow. In the second, the researchers have fenced off cattle from the stream, Gurnsey Creek. In the third, the scientists have planted willows and built beaver dam analogs (BDAs)—human approximations of beaver dams. And last, at the far end of the meadow, beavers have returned and built the real thing.

There are 17,039 meadows in the Sierra Nevada range, just to the south of here, covering 191,900 acres. Childs Meadow is different from these others because it has volcanic soils that are extremely porous. (Mount Lassen is the southernmost Cascade.) Sierra Nevada meadows, on the other hand, have dense granite that groundwater infiltrates less easily. Nevertheless, this study offers some broader implications for alpine meadow management. These ecosystems are critical habitat for two-thirds of species native to California's mountains because they have water in the dry seasons, Yarnell tells me. Species in trouble, such as Cascades frogs and willow flycatchers, have only been seen in beaver-created meadow wetlands.

For water supply too, a healthy alpine meadow is the first step in a healthy downstream. Alpine meadows store less water by far than the vast aquifers in the Central Valley. But the timing is important, according to Yarnell. Snow falling at high altitude melts in spring or summer, wetting meadows and slowing water, extending its release downstream into the later summer.

I eventually come upon three researchers who are part of Yarnell's team, kitted out in chest waders, sunglasses, and hats. Leading the crew is Leah Nagle, then a geospatial specialist at UC Davis. At one site, they've inserted a PVC pipe down into the ground. Into the tube they slip what looks a bit like a microphone. Inside the device are two lines: one positive and one negative. When they hit water underground, the current connects and sends a reading to the attached voltmeter. The whole contraption is attached to a tape measure so they can record the water level in the PVC pipe. They have several such stations set up around the meadow at various distances from the main stream and the four different land uses.

Another test analyzes whether water in a particular spot is primarily snow runoff or coming up from the ground. Water has a geochemical signature that indicates its source. Nagle explains: "Water that has been in the ground for a while picks up mineral traces and salts and that kind of a thing. And so that'll affect how much conductivity or how much electrical current the water will carry." Snowmelt, on the other hand, has very little conductivity. Nagle shows me the meter on her instrument, a multiparameter probe. As expected in June near the stream channel, the water here has a low reading, showing a greater proportion of snowmelt. Higher electrical conductivity readings typically indicate groundwater, although cow patties can add nitrates to the water, also resulting in a high reading.

As Nagle is describing this to me, a big black cow is edging closer to us. "That's Childs," Nagle says. The crew has been working fourteen-hour days here for weeks and has become familiar with the locals. "This is his meadow."

Every time we look back to the instruments, he steps closer. When we look up at him, he stops and stares, like, "What?! Nothing's happening here. You're imagining things."

"We're not feeding you!" Nagle exclaims. Eventually, Childs decides we're just not that interesting and wanders off.

We walk closer to the stream for another test and the land gets wetter. Because I'm on vacation and unprepared, I'm wearing low-rise hiking shoes. As my footsteps push down into the soft ground, water rises up. I'm stepping gingerly, trying to stay dry, but before long, muddy water spills over the top into my shoes. I squish along in the hot sun with a new appreciation for the researchers' chest waders and long sleeves.

Nagle and her team are deploying another instrument, a flow meter, placed a few inches in from the edge of the stream so they can read the rate at which the water is moving. They track the velocity at this same spot (and others) at different times of the year to measure how much water is moving through the channel. Wading into the stream to place the meter, Nagle loses her balance and tips over. The cold water rushes into the top of her chest waders. She laughs, shaking it off. "Wetland work is often . . . wet."

The four-year study, published by the Center for Watershed Sciences, found that, not surprisingly, water levels underground vary with precipitation. Rainier years have higher water levels than drought years. Amazingly, though, the power of beavers overcame those trends. What was "super interesting," Yarnell says, is that although 2015 was a drought year, groundwater levels then were significantly higher near beaver ponds than during 2017, a very wet year that blew out the beavers' dams, dispersing water from their ponds. (The beavers returned and repaired their dams in 2018.) The BDAs, small dams built by humans, also increased groundwater levels. The study found that both human and beaver ponds' influence on groundwater elevations extends approximately thirty-three to sixty-six feet. "It's a local influence only. Ideally, you'd want a distributed network of BDAs or beaver dams to keep surface water there all summer." Just as with other Slow Water projects, giving water space along its path in a series of small areas adds up to a significant impact. It's a different way of thinking about water than we've become accustomed to with our giant, centralized dams.

Beavers in the City

Although we tend to think of beavers as animals who live in rural areas, at one time they lived in the streams and rivers that thread through and under our cities. And now they're baaack! Beavers don't seem to be terribly deterred by our activities. Dittbrenner says he's seen them in ponds abutting a Wal-Mart parking lot, or within fifty feet of a busy highway. Beavers have even returned to the Bronx River in New York City, where they hadn't been seen since colonial times. (A community contest dubbed one of them Justin Beaver.)

Dittbrenner and Bailey studied how many beavers had moved back into Seattle waterways. They looked for streams that meet general habitat criteria for beavers: a slope of less than 5 percent and not terribly incised. They found beavers or evidence of their activity in every suitable stream: fifty-two active and abandoned colonies. "We were really surprised," Dittbrenner tells me. "We thought we were going to see presence here and there, but they were everywhere." Since then, he's heard of numerous sightings around Lake Washington and thinks his findings may be low.

Most urban places where people have allowed waterways to remain on the surface are in parks. Beavers are typically crepuscular—
active at dawn and dusk—so during our midday visit to a few parks,
we don't see any. But Dittbrenner is a water detective of a different
sort. His beaver-acclimated eyes spot things I don't: scent mounds,
tummy trails, a haphazard pile of sticks or leaves that covers a dam,
a footprint in the mud shaped like a triangle with toes. Beaver chew
marks on trees and sticks look a bit like a pencil sharpened with a
knife, the irregular tapered point a mosaic of tooth prints.

Although urban beavers are typically more constrained than their
country cousins and may not be able to store as much water, their
work can still reduce local flooding, create habitat for other creatures, and reduce streamside erosion by cultivating wetland plants
and slowing water. Reducing erosion is particularly valuable in a city
because sediment clogs water pipes and treatment facilities and is
expensive to clean out.

How do we prepare for beavers returning to cities, centuries after they were removed? One approach is to build today for peace
tomorrow by identifying beavery places where people might toler-
ate their presence. "If we anticipate beavers coming, we can build
park amenities and landscape features that allow beavers to improve
systems, rather than create problems," says Dittbrenner. Strategies
include leaving flexible space around water and building elevated
walkways for humans.

Beavers in the wild typically induce a succession regime: they
build dams that create ponds and cut streamside vegetation. When
they consume all the nearby food or the pond silts up, they move
on. Then trees and bushes grow back, and eventually beavers return
to build another pond. An urban park likely doesn't have the space
or human tolerance to accommodate such radical changes throughout the cycle. For example, one of the parks Dittbrenner studied is
designed to use wetlands to clean urban storm runoff, so park managers will likely dredge it periodically so the treatment wetlands
continue to serve. And because people like seeing greenery in the
natural areas they visit, managers might plant additional trees and
other fodder as beavers cut them down. Individual trees that are par-

ticularly special to people are protected with wire mesh or sanded paint around their trunks to deter gnawing.

*

On the Tulalip Reservation, three beavers from another family are ready to move to their new home. Alves, Bailey, and interns load the cages into a truck for the two-hour drive into Mount Baker Snoqualmie National Forest. As we turn onto a dirt road and enter the forest, western hemlock and Douglas fir tower above us, the understory bedecked with western sword ferns, salmonberry, and native blackberry. Along the stream are cottonwoods, alders, and willows.

Although beaver believers spend a lot of time trying to figure out what beavers want and select release sites accordingly, Alves admits, "I will never claim to be able to think like a beaver. If they don't like it, they won't stay." If a stream is deeply incised, a beaver family may work on it for a bit and then give up and move on. Sometimes, after a few families take a whack at it, the site will become healthy enough to entice a family to stay. In other cases, people build BDAs to get the party started, to entice beavers to move in and take over.

When we get to the release site for this family, everyone dons hip or chest waders. Two people carry a beaver in a large rectangular cage. They bring the first one to the stream and set the cage in the water so the beaver can chill after the long truck ride, while the people go back for the second and third beavers. Alves lines up the three cages in front of a "starter lodge" the humans built, which is basically a standing cone of logs assembled in the water with a small opening for the released beavers to squeeze through. The point of the lodge is to give the beavers a safe space to decompress from the journey and to hide from predators. "But we've never had beavers use that lodge structure after we relocated them," Alves muses. "I think it mostly makes us feel better."

She opens the first cage. The beaver hesitates, then walks slowly out into the dark opening ahead, flat tail dragging behind. Then she disappears into the water. In a flash, someone has moved the empty cage out of the way and lined up the second beaver, who follows suit.

Then it's the third beaver's turn. It's all over in ninety seconds. Alves puts two more logs over the opening the beavers just went through, so the only exit is into the water, the way beavers construct their own lodges for safety.

Usually, the first thing released beavers do upon arriving is go get some food, she explains. "After the journey, they're usually pretty hungry. They stress-eat, just like humans." Searching for food is also a good incentive to explore the habitat. The key to natural water management "is people just backing off, you know?" Alves says earnestly. Expensive engineering projects have similar restoration goals, but they're not as effective as "just picking up the beavers and putting them strategically on the landscape."

Beavers are not known for their vocalizations; they're usually fairly taciturn, Dittbrenner tells me. But he remembers one family who seemed to celebrate after their release: "We were getting ready to walk away and you could hear them all squeaking to each other. And it was clear to me that it was a happy, relieved beaver squeak, like, 'Oh, we're back in nature and we're all here! The alien abduction is complete!'"

· 5 ·

Reclaiming Historic
Water Knowledge in
Modern India

From a minivan on the shoulder of Old Mahabalipuram Road on the south side of Chennai, hemmed in by honking trucks and autorickshaws, I watch a painted stork move with quiet dignity through the long grasses of Pallikaranai Marsh. With each step, knee flexing backward, the webbed foot closes, then spreads open again to find purchase on the soft land. As it tips toward a fish, white-and-black-striped tail feathers spread, flashing a surprising red whoosh. Nearby, an endangered spot-billed pelican swirls in for a landing, green-backed herons fish, and gray-headed swamphens tend to their young among cattails and sedges. I am watching from the vehicle because, with the traffic hurtling by, it's not safe to get out. Despite of the calm beauty of the birds—just a few of the 349 species of flora and fauna found here—I feel claustrophobic. For myself, but more so for this delicate ecosystem hemmed in by development. Just across the marsh, a network of power lines, buildings, and roads stretches beyond my view.

Chennai lies on the southeast coast of the Indian subcontinent, and the natural landscape on which it was built is particularly rich in water. Pallikaranai Marsh is linked hydrologically with a complex system of three rivers—the Adyar, the Cooum, and the Kosasithalaya—as well as backwaters, coastal estuaries, mangrove forests, and ancient human-built lakes. This slow water masterpiece once covered seventy-two square miles. Acting like a sponge, it held rainwater, then released it slowly, the water creeping from fresh to brackish to salt, ultimately exiting into the Indian Ocean.

But in recent decades, Chennai has sprawled into India's fifteenth-largest city by area, from 18.5 square miles in 1980 to more than 165 square miles today. The growth has cost the waterlands. An assessment by a local NGO, Care Earth Trust, found that Chennai had lost 62 percent of its wetlands between 1980 and 2010. Pallikaranai Marsh has been literally decimated, losing 90 percent of its area to malls, restaurants, hotels, hospitals, and information technology firms. Just 2.4 square miles remain. It's a problem common around the world. The relatively new IT corridor here is an echo of California's Silicon Valley, where Google and Facebook squat on filled-in marsh.

Nonhuman lives have less and less space to exist across India, where 1.4 billion people jostle to survive and thrive in a land area one-third the size of the United States. But this isn't a story of people versus wildlife. Because when wildlife suffers, people do too. The destruction of Pallikaranai Marsh and other local wetlands has not just made wildlife homeless. It has also spawned dueling water problems—both flooding and shortages—for the people of Chennai. These bifurcated water disasters are all the more tragic because early Tamil people, whose cultural and linguistic heritage continues proudly in today's residents, had developed an elegant system for capturing and holding the rain that fell during monsoons, saving it for the dry season. Their method also replenished groundwater and minimized erosion from the heavy rains. And it supported rather than devastated wetland habitats.

Ancient Tamil water techniques are of a piece with other early cultures that innovated ways to live within their local water means at a time when the world was less globalized. In southern Africa's Kalahari Desert, the San people went crepuscular, limiting their activity to dawn and dusk, staying still in the shade during the scorching midday, even breathing through their noses to avoid moisture loss through the mouth. Nabataeans in Jordan's Petra carved troughs along the red rock walls of the Siq (which had a starring role in *Indiana Jones and the Last Crusade*) to collect rainfall and funnel it to cisterns. In the Andes Mountains of Peru, farmers directed high flows from winter rains underground, to slow the water way down and save it for the dry season (more on that in chapter 6). Across India, people came up with various methods that fit with their local climate, ecology, and geology. These ancient techniques aren't entirely lost. People still practice them in some places, including here in South India, where remnants of the Tamil system remain.

Today, a loose team of people in government, academia, and NGOs in Chennai are working to restore natural and human-built systems to soften water's peaks and valleys, while reconnecting locals with their heritage and holding space for other animals.

*

Chennai's natural water richness makes what happened in summer 2019 all the more shocking: the city grabbed international headlines when it ran out of water. Government trucks made water deliveries to roadside tanks, where people queued with vessels and occasionally brawled, resulting in at least one death. When I visited in mid-November of that year, water trucks still plied the streets.

But 2019 wasn't an anomaly, and Chennaiites were less surprised than the wider world. Over the past two decades, Chennai has regularly run out of water during summer months. That's because paved surfaces throughout the city prevent rain from being absorbed and replenishing groundwater that could be used during the dry season. Balaji Narasimhan, a professor of engineering who specializes in hydrology at the Indian Institute of Technology Madras, sits down with me at his office to explain the situation. Simply put, Chennai shouldn't be running out of water at all. During its few months of monsoon, the city actually receives one-and-a-half times more rainfall than it consumes annually. But today's water managers rush that rain away in stormwater drains and canals, moving it rapidly out to sea. When they need water later, they turn to dwindling groundwater, distant supplies, and desalination plants.

Among Chennai residents, the more emotionally and politically destabilizing event than the water shortage in 2019 was the 2015 flood that killed at least 470 people, displaced hundreds of thou-

FIGURE 5.1 Water trucks such as this one ply the streets of Chennai and other cities throughout India and worldwide, delivering water to people who are unconnected to city water or when the city cannot deliver. In Chennai, Nairobi, and elsewhere, truck water is often groundwater.
Photo © Erica Gies

sands, and left many stranded in their homes for weeks. My friend Uttara Bharath lives with three generations of her family in a building designed by her architect parents in the Saidapet neighborhood. Their home is not far from the Adyar River, which winds through the center of the city, and her apartment is on the ground floor.

While visiting, I asked about their experiences in the flood. During the night, water rose steadily in their home, reaching five feet in their living room and kitchen. Uttara's daughter, Anya Kumar, was twelve at the time. "It was honestly very bizarre to see our furniture floating in the murky water," she tells me. As Anya walked across the wooden floorboards, "they started rising underneath my feet. I sort of floated my way to the kitchen to see what we could salvage."

Uttara's mother, Jayashree Bharath, was surprised to see a large SUV floating in front of their house. "A lot of defense people were swimming up and down on the roads, trying to save people," she recalls. "Whenever they felt tired, they used to get up on the SUV and sit there." Also rescuing people were members of the fishing community, who had boats.

As the recovery got underway, so came the reckoning. The family's cleanup took more than a year and demolished their savings. For people of lesser means, who lost everything and had no reserves, it was even more traumatizing. The water had come in surprisingly fast. A lot of people who lived closer to the river "didn't have any time to run away," Uttara's mother says sadly. "Still, to this day, there's not been a proper count of the number of casualties."

Ironically, it was likely the local mindset of water scarcity that made the flood more deadly. As writer Krupa Ge documents in her book about the flood, *Rivers Remember: #ChennaiRains and the Shocking Truth of a Manmade Flood*, reservoir managers were reluctant to release stored water ahead of the monsoon rains. When they finally recognized the threat, they discharged too much, too fast.

Although the 2015 flood was extreme, it's not uncommon for rains to flood vast swaths of Chennai. The sprawling pavement that prevents water storage underground also pools heavy rain. The city has seen increasingly frequent and intense cycles of both flooding and drought over the past two decades. Although it may sound bizarre for a place to suffer both flooding and water scarcity, a growing number of urban areas around the world, such as Mexico City and Beijing (more on the latter in chapter 7), are experiencing similar problems. Poorly planned development, amplified by climate change, is exacerbating water extremes.

As a moderate rain fell when I was there in early December and streets began to flood, one local aptly captured Chennai's dysfunctional relationship with water in a tweet: "till last week, the residents were booking water tankers and from today they will book rescue boats. What a city!"

Protecting What Remains

Beneath the peepal and tamarind trees, among the flower stalls and idli restaurants, greater Chennai's eleven million people go about their business. Some men wear traditional *lungis* (others wear button-down shirts and pants), and women are dressed in vividly colored *sarees* or *salwar kameez*, with strings of fragrant frangipani in their plaited hair. Cows and dogs wander and nap at will, while jungle crows, magpies, and dragonflies swirl above the fray. Chennai is more chill than the northern megalopolises Delhi and Mumbai, but it shares that quintessentially Indian sheen of chaos that, upon longer observation, reveals an innate order. An unspoken dialogue of push and pull among countless beings following their individual paths somehow manages to keep the whole in a constant state of flow.

Now people are freshly motivated to once again include water in that flow. The 2015 flood seems to have been a turning point that led both the public and the city government, called the Greater Chennai Corporation, to recognize that Chennai needs to change its relationship with water. There's a growing understanding of the role that poor development planning plays in making water shocks worse, and a dawning realization that water scarcity, flooding, pollution, and groundwater recharge are all connected. One government official makes that link clear: "We do not want to compromise our infrastructure development by sacrificing or destroying or degrading the nature," he tells me. After the flood, the Dutch Office of International Water Affairs advised officials on recovery, underscoring that message of holistic water management. The following year, in 2016, the Dutch engaged local government officials, water experts, NGOs, and communities in a multiyear design-and-development program called Water as Leverage.

Together they produced reports that linked existing projects and laid out new ones that would conserve and restore natural and human-built water systems across the entire watershed. The ambitious plans would slow the flow of water to soften the peaks of floods and droughts by reclaiming floodplains, protecting remnants of marshes, restoring ancient human-built water systems, and reconnecting these disjointed places to give water a human-friendly path to flow. As with the other Slow Water projects we've seen so far, doing so requires many small projects rather than standard development's few big ones. To get a sense of what this looks like within a densely inhabited city, I visited numerous projects across Chennai.

An obvious step is to protect any remaining waterlands, such as the remnant of Pallikaranai Marsh I visited. The fact that any natural water arteries still remain in Chennai is thanks in significant part to Jayshree Vencatesan, a biologist who founded Care Earth Trust in 2001 to protect the marsh and other bodies of water around Chennai.

Vencatesan thwarts cultural norms pushed on her gender almost as a raison d'être and is used to being the only woman in the room. Yet she's weary of men telling her, with a mixture of admiration and disapproval, "You're not a normal woman." During early field-work for her doctorate, she tells me that she would booby-trap her sleeping quarters to protect herself from harassment or worse. At her office not far from Adambakkam Lake, pictures of her beloved, deceased Dalmatian along with posters of local birds decorate the walls. Dressed in a *salwar kameez*, her long hair pulled back in a traditional braid, she is the undisputed leader of the organization. Our interview is interrupted when an employee brings in his young daughter for a traditional blessing from Vencatesan. She seems a lit-tle embarrassed but also flattered. Nevertheless, she is not a person who lets others' perceptions of her—pro or con—slow her down. When she began her conservation work, "people said it was the stu-pidest thing anyone could do," she laughs. "But if people challenge me, saying, 'you cannot do a bit of work,' I will take it up."

Vencatesan documented the historical cascading system of sixty-one wetlands and water bodies across the watershed that drain into Pallikaranai Marsh and juxtaposed them with time-series maps showing what's been lost. Catalyzing public awareness, her findings were the basis for a ruling by the Madras High Court to prohibit fur-ther encroachment on wetlands by development. It also called for a state plan to restore some of these ecosystems to reduce city flood-ing and slow water on the land, giving it time to recharge aquifers. National regulators took note, calling out the state of Tamil Nadu for endangering national marshlands and water bodies. The result was a mandate and a budget to improve the situation, including protecting and restoring the city's remaining water bodies.

Vencatesan and Care Earth Trust have been closely involved with the Dutch–local Water as Leverage initiative. Initially, "the government was amused" by the groups' presentations, she remembers, given the officials' general bias in favor of standard development approaches, such as desalination plants, levees, and filling in wetlands to "reclaim" land. Even now the government still plans to build two new desalination plants and a distant reservoir in a bid to reduce water scarcity. But when officials reviewed the final proposal, they were impressed by the water detectives' understanding about the city and its hydrology, Vencatesan says proudly. This initiative and the court ruling have helped put the city on course for change: "Until now, nature has been treated in Chennai as an externality, never factored into urban planning." As this revolutionary shift takes shape, Vencatesan predicts that sand dunes, marshes and other wetlands, and remnant patches of dry forest will once again become "the natural buffers to the city's shocks."

Back to the Future: The *Eris* System

To make that course change, people can look to how their ancestors worked with nature to finesse water cycles. Starting at least two thousand years ago, ancient Tamil people ensured that they had water year-round by building a series of connected ponds that run from the Eastern Ghats—a mountain range forming a north–south spine down the middle of the subcontinent— downslope east to the Bay of Bengal. These *eris* (the Tamil word for "tanks") are open on the higher side to catch water flowing downhill, while the lower side is closed with an earthen wall called a bund. An overflow divet in the top of the bund gives excess water a path to continue on to the next *eri* downhill. "System *eris*" were built off rivers and creeks to capture their peak flows, while "non-system *eris*" were dug in areas without natural waterways to capture rainfall in a series of connected depressions. *Eris* were described in early Tamil literature and temple engravings, says enthusiastic amateur historian Krishnakumar TK, whose day job is in information technology.

To learn more about the history of the water bodies, I visit Krishnakumar, who goes by KK, at his apartment in a newish development on the southwest outskirts of town. He invites me in, gesturing for me to leave my sandals outside, in the Asian custom. His elderly mother comes in to listen but soon lies down and dozes off. A few years ago, as KK saw traditional water bodies in Chennai disappearing to development, he began to research and map them, soon expanding his work to include those already gone. He blogs about them and takes people on walking tours to explain what's lost, like Joel Pomerantz with his Seep City outings in San Francisco.

The *eris* system is the opposite of modern development's tendency to move water off the land as fast as possible. The early Tamils understood that by slowing water's flow, the *eris* reduced flood peaks and prevented soil erosion. Most important, the tanks gave water time to seep underground, filtering it and keeping the water table within reach of wells. Because they were connected to the water table, the *eris* also served as visual indicators of water availability, says Vencatesan. Seeing the water level in a pond signaled to farmers when to sow their crops and when they need to conserve. Tanks were also part of every temple complex, bringing water into the heart of religion and culture. Rituals and rules dictated system maintenance and water sharing.

The *eris* were not just reservoirs for irrigation; they became part of the region's hydrology and ecology. Because many were connected to creeks, rivers, coastal wetlands, and freshwater marshes, they provided natural waterways their due along the way. Even *eris* not directly connected to rivers helped feed the local hydrology because groundwater systems are extensive, so water absorbed in one place could feed a river some distance away. In fact, the words *lake*, *tank*, and *water body* are interchangeable here because, after so many generations, no one remembers whether a particular water body is natural or human made.

British engineers in the nineteenth century were gobsmacked by the scale of the *eris* system—reportedly more than fifty-three thousand bodies of water across southern India—and the deep knowledge of topography and hydrology required to build it. Alas, British respect had limits. Its centralized management supplanted the traditional system by which villages managed their own local *eris*, removing the accumulated silt each year and using it to fertilize fields. The British neglected this maintenance, and the *eris* fell into disrepair, making it easier to justify filling them in and building on top of them—sadly, a pattern that continued after independence.

As they built roads, the British obliterated the flow pathways that had linked water bodies, says KK, giving the rainwater nowhere to go. "They did not understand our system." Today, many famous city landmarks and neighborhoods—Loyola College, Central Chennai Rail Station, T. Nagar, Nungambakkam—sit atop former tanks and lakes. KK uses old government and British maps in his research, but in some cases, he doesn't need to. "My mother, who is eighty-one years old, she would have seen Nungambakkam Lake when she was twelve or thirteen," he tells me. Street names such as Spur Tank Road and Lake View Road commemorate ghost water bodies that once sustained and protected their neighborhoods. Fewer than one-third of the 650 water bodies that KK has documented in and around Chennai remain, decreasing the surface area of water to one-fifth its 1893 extent. And Chennai is not alone in this: Bengaluru, India's booming technology capital in the neighboring state of Karnataka, has followed a similar development path, filling in its *eris* and creating parallel problems with water.

Ironically, given KK's passion for finding and documenting historic water bodies, the IT company he works for is in a special economic zone built atop Pallikaranai Marsh and the neighboring Perumbakkam wetland. He chuckles ruefully, showing me the area on a map. "We used to have hundreds of thousands of migratory birds visiting this marshland some twenty years ago. I have seen [it] getting destroyed in front of my eyes." With just 10 percent of the marsh area remaining, he says, "even I can't [see] the water from my workstation, and I'm on the fourth floor." But as the title of Krupa Ge's book invokes, rivers remember. This area hasn't forgotten it is a marsh. During the 2015 monsoon, it flooded to the second story.

One place KK can find water is right outside the front door of his apartment. What appears to be a lake is lapping up against the building at the time of my visit. Ducks swim by, their quacking juxtaposed against the ever-present car horns that Indians wield as accident insurance. In fact, this is not a lake but farmland flooded by recent rains. It's a small echo of 2015, when flooding prevented KK from leaving his home for almost a month. "There was no electricity. There was no drinking water. There was no telephone connection, nothing." But, he adds, "I was fortunate because I did not have all these buildings in those days," gesturing toward neighboring apartments, so water receded to those areas.

"You mean those buildings have gone up in the last four years, since the flood?" I ask in disbelief. That means a 2015-level flood now would be even worse for him.

"Yeah." KK's tone is matter-of-fact. "That, you cannot avoid."

Wetlands = Wastelands

Back at the Care Earth Trust office, Vencatesan points out another British legacy that facilitated the destruction of wetlands: their official designation as wastelands. To her, the notion of wetlands as wastelands is anathema. "I grew up in the hinterlands, where this notion of waste doesn't exist. To us, nothing is a waste." That attitude was once widely shared across southern India. The fact that the "wasteland" label still applies makes Care Earth Trust's success in conserving some areas all the more remarkable.

Many areas the British saw as waste had been previously designated as shared-use commons, called *poromboke* in Tamil and dating back to medieval times. The ethics surrounding the use of the commons are even older, Vencatesan tells me, rooted in Tamil scriptures. They describe the resources that wetlands provide—fish, seasonal agriculture, grass to weave mats, fodder for animals, medicinal plants—helping people to understand that wetlands and other ecosystems are multifunctional habitats, supporting not just humans but other species as well. They also make clear the requirement to protect wetlands, including penalties for those who don't. As a common property resource, water was subject to rules regarding how it was allowed to overflow from one water body or wetland to another. "This is essentially upstream-downstream equity, you know?" Vencatesan says.

As she learned more about wetlands throughout her career—in part by working with agrarian groups who continue to live close to the land and water—Vencatesan internalized these values of multi-purpose landscapes. She also learned that it's critical to allow certain wetlands to follow their natural rhythm and go dry part of the year to support the life cycles of animals and plants, including crops. "All of our melons and gourds and stuff like that used to be grown when the moisture is retained but the surface flow is not there."

Since the British viewed land as property, these commons, which could not be bought, sold, or built upon, "presented a very peculiar problem" for the colonizers, says Nityanand Jayaraman, a community activist with a collective called Vettiver Koottamaippu. "From a revenue point of view, it was wasteland." Jayaraman works with people in North Chennai, where industrial facilities like coal plants are displacing fishing communities. He also advised the Water as Leverage project. In shorts and a T-shirt, with graying, shoulder-length curly hair, he sits cross-legged in his tiny office, his quiet-yet-fierce voice sometimes drowned out by the enthusiastic roar just outside from the youth activists he's training. As lands surrounding *poromboke* areas were developed, tension mounted over these two sets of competing values. "Of course, the old values lost," Jayaraman concludes. "And what we have is a disaster called Chennai."

The lost values are arguably as significant as the declines in sustainable subsistence and healthy, functioning ecosystems; and, of course, they are linked. Because people's identities are entwined with their land, when development annihilates a place's natural heritage, people also suffer deep cultural loss, a loss of identity. For example, Pallikaranai Marsh is home to neithal—an endemic, striking, blue-violet water lily, one of the earliest flowers described in Tamil literature. Other beloved creatures of the swamp include the glossy ibis and, perhaps surprisingly, the venomous hump-nosed viper. "Snakes are revered in Tamil Nadu," says Vencatesan, who grew up in a neighboring state.

In North Chennai, parts of which remain somewhat rural, Jayaraman has witnessed that loss in a single generation. "Among the older people, there is a far more intimate knowledge of hydrology, of seasons," he laments. "Among the younger people, that is eroding quite quickly. It erodes with the landscape. Your culture also goes with the landscape."

With Chennai's growing population, much of the remaining *poromboke* lands along water bodies and rivers have been occupied by marginalized people. During the 2015 floods, people living along the Cooum and Adyar Rivers were among the hardest hit—but their homes were also part of the problem, hardening embankments and floodplains, leaving no open land for the excess water.

In a controversial decision, local and state government agencies relocated tens of thousands of people from river banks and canals near the city center to newly built high-rise tenement buildings on the city's southern edge. Close to the IT corridor, the tenements are in the Perumbakkam neighborhood, the name commemorating the Perumbakkam wetlands it sits atop. "In its infinite wisdom," Jayaraman says sarcastically, to protect people from flooding, the government moved them from a river floodplain into a wetland just zero to two feet above sea level. It was an out-of-sight/out-of-mind decision that completely ignored the reality of local hydrology. For that reason, the move failed to make vulnerable people safer, Jayaraman concludes.

Instead, "greening the Cooum [River] has become an excuse to evict people we don't like, and those are Dalits and Muslim minorities who live on the river banks because the city has not provided them with a decent place to live," he continues. Because their new residences are about fifty minutes by car from their former homes, many of them have effectively lost their jobs too. And the city's actions have actually made all of Chennai more vulnerable by filling in even more wetlands for the new developments.

The government has also begun to evict and resettle thousands of families from the banks of the Adyar River. But this time, it has vowed to take a different approach. Called proximity relocation, the idea is to redevelop homes in place or to find areas for people to live near their former homes so they can remain within their neighborhoods and communities.

The architecture firm Madras Terrace, which has experience in nature-based water management, is working with Water as Leverage

to plan for proximity relocation in Chitra Nagar, a neighborhood
of corrugated tin-and-wood shacks and tenement buildings that
abuts the Adyar. On another day I visit the area with Sudhee NK,
an engineer and financial planner who works for the firm. We turn
off Anna Salai, a main thoroughfare in the middle of the city, onto
a dirt road, where goats, chickens, and dogs wander about. During
the 2015 floods, this neighborhood flooded as high as two stories,
Sudhee tells me, shaking his head. The architecture firm is proposing to redevelop rather than relocate this community. The plan calls
for restored, taller apartment buildings next to treatment wetlands
that would use plants and natural processes to clean the buildings'
sewage. It would also expand the floodplain along the riverbanks,
reclaiming space to detain water and buffer human neighbors when
the next flood comes.

Chennai's Citizens: Water Warriors

In a way, modern Chennai residents are much more savvy about water than most North Americans or Europeans because clean water
doesn't automatically flow from their taps. As in most Indian cities,
the piped city water supply is available for just a few hours a day or
every few days. People must plan to obtain the water they need. My
hotel room has a bucket and scoop in the shower because the idea of
just letting a shower run is "sacrilegious," explains my friend Uttara.

I learn more about this daily reality when I meet up with Naaz
Gani, a young journalist who writes for the New Indian Express and
lives in a neighborhood called Gopalapuram. Water is delivered to
her apartment building in the middle of the night, she tells me, and
a building manager pumps it to the roof to establish water pressure.
It is available to residents from 7:30 to 9:30 a.m. She runs the tap and
fills a bucket or two for her daily bathing, dishwashing, and toilet
flushing. But if she doesn't get up at 7:30, or if she has to be at work
early . . . "Too bad. You ask your friend to do it for you." Entertaining
guests is also difficult. When her parents visited, she had to borrow
extra buckets from friends.

But the 29 percent of the population who lives in informal settlements without city connections has it much worse, Gani reminds me. People queue with plastic buckets to take water from government tanks or from public groundwater pumps on the side of the road, rolling it back to their homes on hand-pulled trolleys. Still more problematic, city water is not considered safe for drinking. If people can afford it, they either run the city water through a purifier or buy treated drinking water.

Uttara's family, like many who own property here, dug a borewell to supply additional water when shallower rain-fed wells run dry. If the borewell also runs dry, they buy tanker water to store in their cistern, "which is definitely something that only some people can afford," she acknowledges.

It's not just poor people who lack city water connections, a water utility official told me. Many newer areas of the city, especially to the south, are also unconnected. Residents there buy tankers of water. Almost fifty-three million gallons daily are shipped around the city. Water served up in tanker trucks is often extracted from underground. But groundwater pumping here is unsustainable because rainwater cannot percolate back underground where it's blocked by pavement. With few restrictions on use, Chennai's groundwater table is dropping about four to eight inches every year.

And just as the Pajaro Valley farmers in California experienced, pumping out groundwater near the ocean creates a pressure vacuum that lures in seawater, turning aquifers salty. Long-term residents here have seen this problem firsthand, and one of them was inspired to do something about the problem. I meet Sekhar Raghavan, known as the Rain Man, when I go to the Rain Centre, a showplace of rain-harvesting methods just south of the Adyar River. For Raghavan, it all started when he moved to the beguiling beachside neighborhood of Besant Nagar in 1970, shortly after it was developed. In front of laid-back coffeehouses and diverse restaurants, the wide beach is often the scene of carnival games, music, and other events. When Raghavan moved in, the beach was still the domain of fishers. Monsoon rains absorbed rapidly into the sandy ground, and open wells filled up nearly to the surface. Sometimes fishers would draw the sweet water to sell door to door.

But soon, as new developments spread more pavement, Ragha-van noticed the water table was dropping and Besant Nagar's for-merly sweet water was turning brackish. A small, older man with big hand gestures and rectangular black glasses framing sharp, yet kind eyes, he recounts his realization at the time: "The only solution is to push rainwater into the soil." Raghavan founded the Rain Centre to educate others how to do this. He shows me a plexiglass replica of the building we're in, then pours a glass of water onto its roof. Its runoff drains to one corner, through pipes, and down into the well and cistern. In 2002 he convinced the state's chief minister to mandate rainwater harvesting at every house. Despite that success, a Rain Centre survey a few years ago revealed a compliance rate of only about 40 percent. But the 2019 water shortage ramped up inter-est again, he says with a slight air of vindication.

Narasimhan, the hydrology professor, is not surprised at the low rate of compliance. Not all geology is ideal for storing water under-ground, he points out. About 70 percent of Chennai overlies rock or heavy clay, which water cannot move through easily. But in such places, people can drill down through the impermeable layer to move water underground. Both the city government and private in-dividuals are now digging these recharge wells.

Outside the Rain Centre on the day I visit, four men are digging a recharge well by hand. Barefoot, wearing traditional *lungis* and button-down shirts, two guys at the bottom of a deep hole shovel dirt into a shallow basket. Two others hoist it up with rope, while a fifth squats on the rim, watching. They must dig about fifteen feet down to reach porous geology here, Raghavan tells me. The wells have a diameter of three to five feet, depending on the extent of the catchment area they're meant to drain, from one hundred to three hundred feet. Erosion is kept at bay with cement rings that line the well, separated by one-foot gaps to allow some water to seep into the ground through the sides. But most infiltration happens at the bottom. The width also allows people to descend into the well peri-odically to clean it of sediment and garbage. In Chennai today, these wells are becoming more frequent sights, their round perforated covers pocking city streets and sidewalks.

Restoring Temple Tanks

Through Water as Leverage, Madras Terrace has proposed another approach for getting water into the ground across the city: using a remnant of the *eris* system, the temple tanks. KK told me that in the past, every village had a temple, and every temple had a water body. Today, many of those villages, temples, and water bodies have been subsumed by the city.

One bright, blue-sky day, I meet up again with Sudhee NK so he can explain this project to find space for Slow Water within the city. We rendezvous in charming, bustling Mylapore, a neighborhood centered around Kapaleeswarar Temple, marked by a 120-foot-high, pyramidal tower intricately carved and painted brightly with some of Hinduism's more than three thousand gods. Vendors sell flower offerings, small deities, and unglazed clay cups traditionally broken after drinking. One of Chennai's most notable tanks sits alongside the temple, occupying a city block. With its top at street level, an inverted, stepped pyramid descends into the ground so people can continue to access water as the table falls.

Sudhee is wearing a purple linen kurta, white pants, and sandals, and as he guides me around, he is frequently on his phone, texting. Yet somehow he effortlessly steps over cow patties, for which my sandals seem to have a homing device.

"How do you do it?" I asked him, marveling.

Glancing up from his phone, he laughs briefly. "Years of practice."

Sudhee explains that, historically, the tank bottoms were unpaved, so they were connected to the groundwater system. Underground water replenished the tank from below, and rain and runoff from above helped recharge the water table. Temple tanks were connected to larger *eris* systems and served ritual purposes as well. Today the Kapaleeswarar tank holds water—ducks swim along the side and turtles bask in the sun—but only because the bottom was paved about ten years ago to retain water for religious ceremonies. This water is effectively a mirage.

To show me the true status of the water table here, Sudhee leads me across the street to another temple tank, Chitrakulam Pond, be-

FIGURE 5.2 Kapaleeswarar Temple tank in the Mylapore neighborhood of Chennai covers a city block. It has been paved at the bottom so water is permanently available for ceremonies, but local water detectives want to remove the concrete, restoring it and other tanks across Chennai to their functional purpose as water catchments and groundwater recharge sites.
Photo © Erica Gies

lieved to be more than two thousand years old and not cemented. Its bottom is carpeted with a mat of fresh grass, sprouted from recent rains, but the water has descended deep underground. "This is the real situation," Sudhee bemoans. Too much pavement and too many borewells are to blame for water levels more than sixty feet below the surface.

Sudhee and his colleagues at Madras Terrace want to restore temple tanks across Chennai to their natural, unpaved state to move water underground. The city government is connecting stormwater drains to temple tanks wherever possible to allow for groundwater recharge. And Sudhee's team is also helping to raise the water table by collecting rainwater from buildings via bioswales, those vegetated ditches, placed wherever they can find room along streets, on

FIGURE 5.3 This temple tank, known as Chitrakulam Pond, shows the real status of the water table: belowground. This tank is just across the street from Kapaleeswarar tank. Photo © Erica Gies

hotel properties, in backyards, and in schoolyards. A pilot project in a neighborhood school is underway, which has the added benefit of educating the next generation about these issues, says Sudhee. However, water doesn't always just infiltrate open space. In another neighborhood, a man pointed to the schoolyard over his fence and told me water pools there when it rains. If soil is compacted, say, from children's feet, or if it's mostly clay, people might need to mix in some sand or build recharge wells to allow water to percolate.

This patchwork of small bits of permeable land can collectively make a big difference. The Mylapore project is expected to provide an average of one million gallons of water per day, nearly half the neighborhood's demand. Replicating the project across fifty-three other temple tanks in Chennai could result in sixteen million gallons per day of recharge, about 6 percent of the city's demand, according to the Water as Leverage team's projections.

Connecting to Flow

Larger areas of open water and wetlands still remain in Chennai, mostly on the south end of the city. Care Earth Trust is working to conserve these elements of the natural water system along with Pallikaranai Marsh. On a visit to a few of these sites with Care Earth Trust staff, I see candleflower—a medicinal plant whose flowers exude a milk that soothes skin injuries—as well as bronze-winged jacanas, fish eagles, black bazas, northern shovelers, and many other species.

At Thalambur Lake south of the city, an area that had been devastated by quarrying, the Care Earth Trust team cleaned and repaired drainage channels that had been clogged and encroached upon. We walk along a bund they shored up to keep water in the lake longer so fish can lay eggs in it once again. Across the bund's surface, they've planted young saplings of peepal and native bamboo to stabilize it. Looking down into the restored lakebed, I see the small hummocks they built and planted with trees to serve as island nesting sites for birds once the water returns. A year after my visit, they invited the water back in. Thalambur Lake now covers seventy-six acres, is protected as a conservation area, and has a laboratory on site for scientists to study its ecology.

At another spot just off Mahabs Highway, south of Pallikaranai, is a dock where people rent pedal boats to explore Muttukadu Backwater, an estuary close to the Indian Ocean that is buffered from full tidal flows by a sandbar. Dozens of pelicans sit on the water, bobbing on its calm surface. Here, as in Besant Nagar and in other neighborhoods near the ocean, excessive groundwater use has allowed seawater to push in. Muttukadu has grown too salty for some of its native fish, and water levels have dropped. In response, Care Earth Trust staff are educating local people, petitioning the government to regulate water extraction, and restoring pathways upstream to allow fresh water to flow into the wetland and replenish it. They are also replanting mangroves to protect the coast, clean the water, and improve breeding habitat for fish.

These myriad projects to reduce droughts and floods, scattered across the broad landscape of the city, are decentralized management, a partial return to the region's traditional ways. Yet despite stated government support for these Slow Water projects, getting utility engineers to embrace green solutions is difficult, mulls Narasimhan, the hydrologic engineer at the university. The systems are more complex than concrete-lined drainage channels, levees, and dams. Slow Water projects usually have a biological component, such as plants that may require soil amendments to achieve the chemistry or filtration they need, he says. Also, because such projects tend to cover a larger area than concrete solutions do, the public is more likely to come in contact with them, so project managers need to cultivate community understanding and support. But that requirement can be a benefit, Narasimhan hurries to add—reconnecting people with their water.

When water "magically" arrived via centralized distribution, people stopped caring for their water bodies. It was a marked change from the ancient Tamils' active stewardship of their local water. "Even two hundred years ago, people used to worship rivers as goddesses," KK had told me. "Because of that, we were preserving water. Now we've lost those cultural values; we forgot." But maybe it's possible to reverse that trend. As government agencies and NGOs reclaim space to reestablish Slow Water, and as people harvest some of their personal water from local supplies, Narasimhan hopes they will become newly motivated to keep them clean and replenished.

One project, also included in the Water as Leverage vision, is directly targeting the next generation. On a warm, partially overcast day in early December, I grab an autorickshaw to Tholkappia Poonga Eco-park, a 58-acre green haven in the heart of the city, near the mouth of the Adyar River. Separated from the beeping traffic outside by concrete walls, the tree canopy is thick here. The peaceful walking paths—earth below and air above—are bustling with butterflies, beetles, blue-and-yellow grasshoppers, crane flies, lizards, lorikeets, and other birds.

Years ago, a tributary of the Adyar River on this site was filled in for development. But instead, this place became a dump, piled high with garbage and human waste and used for illegal activities. Then more than a decade ago, local NGO Pitchandilkulam Forest Consultants and city agency Chennai River Restoration Trust began restoring the area's former river and estuary habitats. The goals were to counteract pollution and biodiversity loss.

My guide here is K. Ilangovan, an ecologist and wetlands specialist who has overseen this project from the beginning. He has planted thousands of the trees himself, including mangroves and 250 other species of plants. (He has a particular fondness for mangroves; some of these trees he planted south of the city actually saved his life during the infamous 2004 tsunami, when he clung to them to avoid being swept out to sea, then waited there for hours until waters receded.) Since the eco-park project began, he's seen snakes, mongooses, mice, and even jackals that feed on crabs and fishes. "We didn't introduce anything here," he says. After replanting the site, "everything came."

People from nearby neighborhoods were involved in the planning and planting so that they would better understand the area's purpose, and some still have jobs here taking care of the plants. During the monsoon, the whole park becomes flooded, slowing and storing water. The surrounding neighborhoods have seen higher water levels in wells and reduced flooding, including during 2015, when they were spared the worst impacts of the devastating flood. The microclimate has also changed, keeping the air a little cooler than the concrete jungle elsewhere in the city. Another phase of restoration has recently been completed, bringing the total to 358 acres.

Aside from offering homes to a variety of creatures in the middle of the city and mitigating water problems, the eco-park is a favorite field-trip destination for schoolchildren, who come here and are blown away. Living in the city with little exposure to nature, "they are so happy to touch and feel the plants," Ilangovan says with obvious emotion. "You can see the brightness in their faces." That outreach—which he also extends his own young son—is critical to changing the direction of society, he believes. "We can't go and change the people with forty or fifty years' age. So I focus more on the kids."

Amid the growing environmental awareness among the general

public and the government, several of the projects proposed by the Dutch–local partnership are moving forward, reports Vencatesan, including the Mylapore tank and Muttukadu Backwaters. A state wetlands panel recently designated four more water bodies in Chennai for protection and is considering another thirty-one. Separately, Vencatesan's nomination of Pallikaranai as a Ramsar site, a wetland of international importance, is under consideration by the central government in Delhi. But development pressure remains intense. "We as a country are trying to become more of a wealthy nation," Vencatesan points out. "And if we assume that people don't want that, we are wrong."

Back at Pallikaranai Marsh, walking down a muddy path at the edge of shallow open water, we pass trees planted by Care Earth Trust, trunks wrapped in cuttings of an unpalatable plant to deter nibbling cattle. Bee-eaters and kingfishers whiz by, and a fan-throated lizard darts under a rock. The sun glints off something, catching my eye, and I lean into a bush to find an iridescent-green jewel beetle with black spots, bumpily navigating stems and leaves. Antennae waving about as if flailing for balance, the insect suddenly flips, revealing a bright-orange undercarriage. Despite the staggering losses in recent decades, there's still plenty to amaze here.

For Vencatesan, educating people about the value of water systems and biodiversity is a long-term process, one that she's been working toward with her characteristic persistence for decades. Near the spot where I had been stuck in the minivan, she is planning to reroute traffic away from the marsh and build a pedestrian "ribbon walk," where people can interface with nature. "They should see a value in it," she says. "Otherwise, it's not going to last."

· 6 ·
Planting Water

HOW WATER SHAPED CULTURE
IN ANCIENT PERU

On a mild day pre-pandemic, during Peru's austral winter, I arise at 5 a.m. to drive north out of Lima, up into the highland village of Huamantanga. I am traveling with scientists who are studying local farmers' use of a 1,400-year-old technique to extend water availability into the region's long dry season. Wending our way through the narrow Chillón River Valley, a slim corridor of irrigated green crops hemmed in by sheer walls of tawny rock, we cross the river and begin to grind up a single-lane dirt road clinging to the side of a steep mountain. In the vertiginous drop-off to our left, vultures sail on thermals at eye level.

The seventy-three-mile drive from sea level to above 11,000 feet is a breathtaking introduction to the country's dramatic terrain. Peru is a land of extremes, about 800 miles long and 350 miles wide. Its location in the tropics near the equator brings heat and humidity to the thick jungle of its interior Amazon. But the shockingly steep Andes Mountains that form the spine of the country hold snow and glaciers, while their western flanks dive down to the arid coastal plain where most Peruvians live. In the mountains, dichotomies rule. Temperatures swing from the oft-cited "summer every day to winter every night." Precipitation is similarly extreme: rain and snow fall during a few months in the southern summer, and the rest of the year is the long dry.

That period without water has been a catalyst for hydrological innovation again and again. In the Peruvian Andes, water shaped culture, making this area one of the six places in the world where complex civilizations emerged (along with Mesopotamia, ancient Egypt, China, the Indus Valley, and Mesoamerica). These early water detectives cultivated deep knowledge of water and the underground, deploying strategies that still astonish—and that some still use. Canal irrigation in northern Peru's Zaña Valley dates back to 4700 BCE, earlier than in China or Mesoamerica. That first foray into water management was followed by those of storied civilizations well known among historians, including Norte Chico, Chavin, Paracas, Moche, Tiahuanaco, Chimú.

The relationship between water and earlier people has lessons for us today. Their hard-won knowledge of how to optimize an erratic water supply could help modern Peru adapt to climate change. One of the most water-insecure countries in the Western Hemisphere, Peru is seeing water scarcity grow worse as a result of climate change and human activities. Its glaciers are disappearing, and people in the mountains told me that the rainy season has already grown shorter during their lifetimes. But as in California, those longer, drier periods come with more intense rainy seasons, offering an opportunity to capture that bounty. At the same time, the country's population has more than tripled since 1960, and land-use change is hindering ecosystems' ability to regulate these cycles naturally. But today, modern Peruvians are starting to redeploy ancient knowledge and to protect natural ecosystems such as high-altitude wetlands in a quest for greater water security. It's one of the world's first efforts to integrate nature into water management on a national scale.

They don't need to look far for inspiration. The water management strategies of relatively recent cultures remain visible today. In Nazca, famous for its mysterious geoglyphs—giant drawings on the parched desert—people still harvest water from 1,500-year-old *puquios*. These are horizontal trenches dug underground in gravelly terrain, where the river disappears beneath the land. Spiral paths carved into the ground, called *ojos* (eyes), allow people to access water in the *puquios*. These access points also alter the atmospheric pressure inside the canals, funneling in wind to push water along like a pump. In its heyday around fifteen hundred to two thousand years ago, the Nazca civilization used this profound knowledge of water to grow crops in the desert. Part of a once-vast system, thirty-six *puquios* are still used today.

Around the same time, the Huari were thriving in the high mountains. They built canals to divert high flows from alpine streams during the rainy season and spread it across porous basins, where the water infiltrated into the ground and later emerged from springs at lower elevations. Because water moves more slowly underground through sediment than it does running on the surface, this strategy allowed people to harvest the water for irrigation in the dry season weeks and months after it fell from the sky. Today, in at least three Andean villages, people continue to use and care for these canals, called *amunas* in Quechua, meaning "to retain."

Fast-forward to the 1400s, and water management was integral to the design of the famous Incan cities of Machu Picchu, Ollantaytambo, and Sacsayhuamán. At Sacsayhuamán, pipes and trenches guide water away from buildings to prevent erosion and to capture water for use. The steep terraces at Machu Picchu and Ollantaytambo contain tiramisulike layers of soil, then gravel, then rock to absorb water where it falls, stanch erosion, and feed groundwater that later emerges in the Rio Urubamba below. As I traveled throughout the Peruvian Andes, I saw similar terraces in small farming communities, planted with varied crops. Rustic types are called *terrazas*, and a slightly more constructed version with stone walls are called *andenes*. People today use these systems to capture water and infiltrate it, slowing water for their crops rather than letting it run down the mountains. The Nazca, Huari, and Incan techniques are all nature based because they use the underground to slow water.

These innovations were largely destroyed by the Spanish colonizers. In the classic 1992 book *Stolen Continents*, author Ronald Wright quotes from a contemporary Indigenous account. Felipe Waman Puma (a.k.a., Felipe Guáman Poma de Alaya) meticulously reported *The First New Chronicle and Good Government* over thirty years, completing it in 1615. Waman Puma writes about the irrigation canals built by "Indians with the greatest skill in the world." With the water they diverted from rivers and springs, they were able to grow enough food to support a large number of people. The innovations allowed them to produce food on all the land "whether jungles, deserts, or the difficult mountains of this realm." Incan kings ordered that this infrastructure should be protected from harm and livestock. But with the arrival of the Spanish, Waman Puma explains, "this law is no longer kept, and so all the fields are ruined for lack of water." The Spanish let their livestock run free, causing great damage. "And they also take the water, and break the irrigation canals, so that they could not be repaired now for any amount of money. . . . And so the Indians abandon their towns."

Most modern Peruvians have a mix of Spanish and Indigenous ancestors, and some descendants have hung on to earlier cultures' knowledge. Luis Acosta is an agricultural engineer and a supervisor at SUNASS (Superintendencia Nacional de Servicios de Saneamiento, or National Superintendency of Sanitation Services), the country's water regulator. Born and raised in the mountains, Acosta serves as a sort of translator to city people, advising them on farming issues and traditional knowledge. "In Peru, the Andes is the backbone from where the water is actually born," he tells me at SUNASS's offices in Lima. "All of our civilizations in that area understood this."

*

Lima, the capital city, is home to about one-third of Peru's population—eleven million people. The cold Humboldt Current traveling the Pacific Ocean just offshore pulls a cool gray fog over the city, like San Francisco in summertime; yet Lima receives virtually no rainfall, just half an inch per year. (Another third or so of Peru's population lives elsewhere along the coast, also with limited rain.) To support that human abundance, Lima relies on three rivers whose headwaters lie in the Andes that rise behind the city. Lima and Peru's other coastal cities share this reliance on mountains for much of their water with many other people around the world. And with the growing global population, around 1.5 billion people could depend on water flowing from mountains by 2050, up from two hundred million in the 1960s. As glaciers and snowpacks are disappearing, this dependence is deeply troubling.

In Lima, the water utility SEDAPAL (Servicio de Agua Potable y Alcantarillado de Lima, or Potable Water and Sewage Service of Lima) is able to deliver water to customers only twenty-one hours a day, according to Ivan Lucich, president of SUNASS, who I met at the regulator's office tower in Lima. And that delivery doesn't include the 1.5 million people who are not connected to city water. Most apartment buildings have water tanks, so middle-class residents don't notice the gap in supply. But if things don't change, Lucich says he expects water delivery to further erode, maybe by an additional hour each year.

Although Lima stores wet-season water from the mountains in reservoirs behind dams and in an aquifer under the city, it doesn't have enough storage to survive a significant drought, and it is unable to fully capture rainy-season surpluses. Two years of low water back to back and "we are fried," Lucich says grimly. Lima's water management will be inadequate as early as 2030, according to a 2019 World Bank report. Human activities are putting further strain on water resources, especially from soil degradation and land use changes that make the soil less able to hold water.

Lucich seems to feel the weight of this responsibility, telling me that these water problems spur him to "light up candles to our saints," an influence from the country's Spanish heritage. "We also turn to our Andean heritage: we go to the spirits in the mountains. Many Peruvians are actually pantheists—we believe in many gods." He adds, "Ecosystems are more than water." Given this worldview and the rich cultural history tied closely to water, the solution Peruvians have come up with makes sense.

National Ambition: Looking to Nature

Desperate for water security, the country's leaders did something radical: several years ago, they passed a series of national laws requiring water utilities to invest a percentage of their customers' payments in "natural infrastructure." These funds—called Mechanismos de Retribucion por Servicios Ecosistemicos (MRSE, or mechanisms of reward for ecosystem services)—go to nature-based water interventions, such as restoring ancient human systems that work with nature; protecting high-altitude wetlands and forests; introducing rotational grazing to restore damaged grasslands; or reverting to traditional livestock, such as alpacas and llamas, who walk and graze more lightly on the land than cows and sheep do.

The goal is not to replace gray infrastructure but to use Slow Water projects to make the dry-season supply more resilient and longer lasting. These laws compel government agencies to invest in protecting and restoring natural infrastructure in the same way they fund built infrastructure such as dams and water treatment facilities. It's a recognition that actions upstream can help ensure the water supply downstream. Before, it was considered a misuse of public funds if utilities invested in the watershed; now it's required.

Yet, despite Peru's forward-thinking policies, putting them into practice has been slow going, due in part to high turnover in government—including six presidents in six years. Another big hurdle, and one that most countries face, is overcoming ingrained practices in the water sector to try something new.

In 2018, Global Affairs Canada and the United States Agency for International Development pledged to invest US$27.5 million over five years to help Peru get its innovative program off the ground. The money went to Forest Trends, an NGO that has been working on nature-based solutions for water in Peru since 2012. The executive director of its Lima office, Fernando Momiy, has championed the idea for almost two decades—in government as the chief of SUNASS, then via Forest Trends. He oversaw early experiments with tariffs on water users, both in the cloud forest town of Moyobamba to reverse deforestation and in Cusco to address water conflicts between country and city people. The NGO's initiative, called Natural Infrastructure for Water Security, aims to provide technical know-how.

Doing much of Forest Trends' day-to-day organizing is deputy director of the project, Gena Gammie, a bright American with a can-do energy. I meet Gammie in Lima at the Forest Trends office, a house in the upscale neighborhood of Miraflores. She originally wanted to be a climate negotiator but grew disillusioned after a global summit, when she realized that humans will not escape catastrophic climate change. That led her to alter course and focus on adaptation. "Water seemed the clearest way that climate was going to impact people and society: longer droughts, heavier floods," she tells me. "Whatever we can do to improve our water management is also going to help us with resilience to climate change."

Growing up on Maui as a *haole* somewhat prepared her for being a *gringa* in Peru, she thinks, instilling a respect for Indigenous knowledge and local culture. In her elementary school classes, Native Hawaiians gave weekly cultural instruction. Relevant to her work today is the Hawaiian tradition of *ahupua'a*, a land division centered around streams and valleys in which the water was managed holistically from mountaintop to ocean. It's a way of looking at water that governments and other institutions worldwide haven't used in generations. Now, with the money invested in Forest Trends, "we have the resources needed to really make this change, but it's a huge responsibility," she says. "We all feel it a lot."

Across the globe, nature-based solutions are still not given much respect in some quarters. One sign of that is the comparatively minimal funding directed to them (as mentioned in chapter 1). And because many early projects have been small scale, people tend to think of them as attractive side features rather than a key tool. It's akin to the long-held attitude toward solar and wind power that is swiftly becoming outdated: they're nice, but were considered incapable of playing a major role in meeting our energy needs. For any radical departure in how we do fundamental things, like supplying energy or water, there's a tipping point to overcoming the entrenched mindset caused by decades or centuries of doing it one way. With the national ambition of its program, Peru has the potential to demonstrate how effective Slow Water solutions can be on the scale of watersheds.

In so doing, it could influence people around the world to rebuild healthy relationships with water based on their own ecologies and local knowledge. Cultures elsewhere have their own histories of collaborating with nature by infiltrating water underground. Peru's *amunas* have parallels in various communities around the world: Paiute canals to recharge aquifers in Southern California; *acequias de careo* in southern Spain; and *qanats* or *karez* in Iran, Iraq, Oman, and elsewhere in the Middle East.

Water and Knowledge Retained

Ancient people's relationship with water is what had brought me on the precipitous journey up into the mountains north of Lima to the village of Huamantanga. The scientists I was traveling with were collecting data on the region's *amunas* to measure the value of this ancient intervention for SEDAPAL and other water utilities. After delaying the water underground, the people of Huamantanga— called *comuneros*, members of an agricultural collective—harvest it to water their crops. Because much of their irrigation soaks into the ground and eventually makes its way back to the rivers that supply Lima, repairing abandoned *amunas* scattered throughout the highlands could extend water into the dry season for city dwellers too.

On the drive up I'm riding with Boris Ochoa-Tocachi, a hydrological engineer who was finishing his doctorate in mountain hydrology at Imperial College in London. He is now chief executive of an Ecuador-based environmental consultancy firm called ATUK and an adviser to Forest Trends. A friendly, loquacious Ecuadorian with some Indigenous roots, Ochoa-Tocachi tells me that, as a child, he traveled through the Andes to villages like Huamantanga with his father, a lawyer who fought for better working conditions for Indigenous farmers. "I had this connection and I really liked the way in which you could see your work having an impact," he tells me with quiet feeling. In college he studied engineering, but was then drawn to follow closer in his father's footsteps when friends—who were fighting oil development in Ecuador on Indigenous lands— exposed him to environmental activism. An unexpected weekend in the highlands with them, he says, "changed my life completely."

As our truck climbs the dizzying mountain, I try not to look toward the gaping chasm to our left. Ochoa-Tocachi, in a gray-and-

black jacket and sunglasses, has made this trip multiple times and casually holds on to the roll bar to brace against the frequent turns. He
wrote his master's thesis on road building, he shares, but when he
came to understand that roads open up the environment to exploitation, "that was a very tough moment for me. I realized that I studied
for six years a career that is mostly about going against nature." That's
why he refocused on hydrology, which he sees as a way to integrate
environmental concerns and his passion for helping rural people.

The *amunas* have also had a big impact in shifting his mindset
from standard engineering. "The first time that I went to Huamantanga, when I saw the canals, it was a revelation," he recalls. "I realized, 'Oh! So we can actually work with nature. We don't have to
build a dam, necessarily, if we want to store water.'" With that epiphany, he became a water detective.

Above eleven thousand feet we reach a plateau where the land
expands around us, revealing fields of avocados, hops, potatoes, and
beans. Finally, we arrive in Huamantanga, where two-story buildings of mud bricks and concrete—some painted, some au naturel—
line narrow dirt streets. Burros, horses, cows, dogs, and people putter around.

We get out of the truck in the central plaza, where the Catholic
church sits with its back to even taller mountains. Ochoa-Tocachi introduces me to one of his colleagues, Katya Perez, now independent
but then a social researcher with the Lima- and Quito-based NGO
CONDESAN (Consorcio para el Desarrollo Sostenible de la Ecorregión Andina, or Consortium for the Sustainable Development of
the Andean Ecoregion). In Huamantanga, Perez cultivated relationships with the *comuneros*, documenting their knowledge and traditions for maintaining the *amunas* and sharing water and communal
work. The social organization is similar to ancient farming communities consisting of kinship groups called *ayllus* that "had been the
building blocks of every Andean state since the empire of the Incas
and before," according to Wright's account in *Stolen Continents*.

Perez takes me to the house of Eugenio Garcia, an older *comunero*
who wears a typical Andean black felt hat, Converse-style canvas
sneakers, and a wide, warm smile that conveys youthfulness despite

the deep lines that crease his face. Huamantanga has very good land but scarce water, Garcia tells me through Perez, who translates. The ancestors' genius was figuring out how to make that work. Confident, a little sassy, he says he's proud to be part of maintaining this tradition.

After others from CONDESAN arrive, we eat a late breakfast at a restaurant that, typical of Andean towns, is in the front room of someone's home. It contains a couple of dining tables and a counter selling a few goods: soap, white melons, cookies. Breakfast is bread and locally made white cheese, with a texture like feta but a milder flavor. A visit to the bathroom leads me down a hall to an open atrium strung with a clothesline, where laundry and drying meat hang side by side.

Popping up from sea level in Lima to above eleven thousand feet in just a few hours means we are not well acclimated to hike to the *amuna*, which sits at nearly thirteen thousand feet. Altitude sickness strikes people seemingly at random; plus, the extremely thin air makes hiking a slow-motion activity. Ochoa-Tocachi and Javier Antiporta, a forestry engineer with CONDESAN, usually hike to the highlands on their trips to collect data. But today, villagers bring us horses to rent. My ride is a black male with a big white blaze who has a propensity for sniffing butts that irritates the other horses. He seems unenthused about the journey, remaining near the back of the pack throughout, but he is gentle and keeps at it. Hauling me up a steep slope, he huffs with effort, and my gratitude and sympathy go out to him.

Some distance above town, we reach another rolling plateau, where the *puna* grassland is scattered with scrubby chamise bushes and lupine in decadent purple flower, and the mountains stack behind each other into seeming infinity. A group of glossy ibis fly by, their feathers glinting iridescent green-black in the strong sun that is quickly making it T-shirt weather, despite the fact that we'll be wearing down jackets and hats by evening. A huge bird wafts high, and I feel giddy from the mix of warmth, natural beauty, and the possibility that I'm under an Andean condor—or maybe from scant oxygen.

In front of us, a rectangular rock emerges from the middle of a small pond. Called a *huanca* in Quechua, the rock extends six feet down, marking it as a sacred area where locals harvest water, Antiporta explains. *Huanca* can be translated as "monolith" or "totem"— basically a sacred marker. Perez tells me later that *huancas* are closely tied to water, and people in Huamantanga consider them to be their ancestors' gods. By putting a *huanca* in a pond, they are honoring tradition and the resource that sustains them. This pond, one of fourteen, collects and holds water after it travels from alpine stream to *amuna*, to infiltration basin, to downhill spring, to stream, to pond. The common name for these ponds, *laguna de los abuelos*, or "grandparents' lagoon," reflects their antiquity.

Overlooking the pond is another *huanca*, a horizontal rock perhaps fifteen feet long, cloaked in lichen and capped with a stone pedestal and a wooden cross. The cross, dressed in a satin scarf, flowers, and tinsel garland, reflects how local people incorporate Catholicism into traditional beliefs, Antiporta tells me. In this case, the crosses are now called *huancas* as well, one religious tradition layered upon another.

Caretaking of resources and each other is fundamental to traditional Andean cultures. In *Stolen Continents*, Wright explains how these societies, "like many others in the Americas, were built on the ethic of reciprocity, not rapacity." The ongoing Indigenous traditions here prescribe these values, including rules that govern *comuneros'* equitable use of water and shared maintenance of the *amunas*. At the start of each rainy season in December or January, the *comuneros* walk up into the mountains for a workday called *faena* ("communal task" in Quechua) to clean debris and silt from the *amunas*. At the end of the rainy season, usually around April, they join together again to clean the lower elevation channels, which have partially filled with sediment while moving the rains downslope.

I learn more after staying overnight in Huamantanga, when I visit *comunero* Ferrer Jimenez de la Rosa at 7:30 a.m., after he milks his cows so his wife can make cheese. The April workday is followed by a party, the Fiesta del Agua, he tells me. Villagers bedeck their hat brims with flowers and make offerings of coca leaves, cigarettes and alcohol to the land and water. Then they reconsecrate the *huancas*,

FIGURE 6.1 *Comunera* Lucila Castillo Flores from the village of Huamantanga told me: "If we plant the water, we can harvest the water." Photo © Erica Gies

decorating the crosses afresh and singing Catholic songs. Finally, they welcome the year's new water manager, a *comunero* who will be responsible for equitably allocating the water and fining those who waste it.

Another *comunera,* Lucila Castillo Flores, a grandmother with a kind smile wearing a skirt and a brimmed hat, explains the party's significance while sitting in her small store off the main plaza with her four-year-old granddaughter: "We believe that it's like a payment to the earth so it can give us enough water permanently."

Indigenous Knowledge Plus Science

In the highlands, we leave the *huancas* and ride further up the mountain, where I finally spy the *amuna* that feeds them. A canal about two feet wide and a couple of feet deep, it winds like a sinuous snake along the contour of the hills for almost a mile. This being mid–dry season, it's nearly empty of water, having delivered its liquid riches to a rocky, bowl-shaped depression where it infiltrated into the ground. Castillo Flores likens what happens here to sowing water, using the verb *sembrar*: to plant. "If we plant the water, we can harvest the water," she tells me. "But if we don't plant the water, then we will have problems."

Made by carefully placing rocks together, the *amunas* are typically not impermeable, and some of the water diverted from the stream leaks out en route to the infiltration basin. The section we have come to see, however, is a 1,640-foot stretch that *comuneros* mortared in 2015. CONDESAN, the Natural Capital Project at Stanford University, and Peruvian NGO Aquafondo financed the work; the goal was to ensure that more of the water moved further from the stream. Ochoa-Tocachi explains, "If water infiltrates too close to the diversion, it will return almost immediately to the stream instead of traveling to the springs." Mortaring is not necessarily required to restore the *amunas*, however, as some of the diverted water will still travel to the infiltration basin.

We dismount, leaving the horses to graze, and walk over to the canal. Just before the start of the *amuna*, where water from a stream is diverted into the channel, the researchers have installed a small weir, much like the ones I saw in Devon, England, at the beaver complex. A classic hydrological tool to monitor stream flow, the weir is a metal plate set vertically across a stream with a V shape

FIGURE 6.2 *Comuneros*, communal farmers in the Andes, use *amunas* like this one to extend their water supply into the dry season.
Photo © Erica Gies

notched into it. It sits between two banks and creates a small pond, raising the water level so that even now, when flow is low, water passes through the V and researchers can measure it accurately. "The flow that you see coming out of the weir is the same flow that is coming into the pond," Ochoa-Tocachi points out to me while the others loll in the sun. "But the structure makes it easy to measure."

The geometry of the V is standard, so the height of the water behind the weir can be converted to volume of stream flow. Here, they measure height with a pressure transducer, an instrument placed in a pipe submerged in the weir's pond. Greater weight on the sensor means higher water and thus more flow.

Above us, sprouting from the cliff that forms one side of the stream's banks, is a plant whose long, bare stalks dangle leis of spikey leaves, reminding me of a midcentury modern light fixture. Each lei ends in a cluster of tubular pink flowers the length of my fingers, with stamens like lilies. Later I discovered this sighting was a brush with celebrity. The plant was *Cantua buxifolia*, sometimes called Inca bells or magic flower. I can see why—it's a botanical wonder. Aside from being gorgeous, it has medicinal, practical, and cultural uses and is the national flower of Peru.

Amid the beauty, however, there is danger. Before leaving the *amuna*, one of our group starts to suffer from altitude sickness, feeling weak and nauseous. Antiporta, prepared for this moment, pulls a small oxygen tank out of his backpack. We sit on the edge of the *amuna*, snacking, while our sick colleague breathes oxygen through a face mask until he recovers enough to continue downslope.

Hopping back on the horses, we ride partway down the mountain and stop at a spring fed by one of the area's *amunas* above. Here, water that's been traveling through rock and soil seeps out into a burbling stream. "You see, it's actually a lot of water compared to the stream that we saw in the weir," Ochoa-Tocachi says, with obvious satisfaction.

One of the most remarkable things about the *amunas* is that villagers know which canal feeds which spring, demonstrating an understanding of the path that water takes underground. By interviewing them, Perez has been able to document their knowledge of these connections, which they have passed down through the generations.

Urbanites tend to discount the expertise of rural and Indigenous people, Ochoa-Tocachi says earnestly, but he and other researchers were able to verify the *comuneros'* knowledge as "very accurate" with hydrological tracking. The researchers added tracers to *amunas'* flows and then used sensitive detectors to track those molecules' emergence in the spring-fed ponds. This finding "surprised us," Ochoa-Tocachi says, and it's important in quieting those urban doubters. "It shows that we can actually use and revalue Indigenous knowledge to complement modern science to provide solutions to current problems."

None of the villagers I meet know how their *abuelos* traced specific subterranean paths between *amuna* and spring. But Garcia, one of the *comuneros* who welcomed me into his home, tells me about another *amuna* they mortared in recent years. Afterward, they observed a particular spring running fuller. It's possible that finding the paths underground may have been the fruit of experimentation paired with close observation.

*

Back in Huamantanga, during a dinner of milk soup, beef curry, cassava, potatoes, and homemade cheese, Ochoa-Tocachi, Antiporta, and Perez explain to me the findings of their study that showed how this ancient knowledge could help with water scarcity in Peru. After measuring flows into this area's *amunas* and out of its springs, the researchers modeled how restoring the many abandoned *amunas* scattered throughout the Andes highlands—and building new ones—could help Lima.

Even with Lima's gray-infrastructure storage of winter river flows in reservoirs, it still comes up about 5 percent short annually, on average. That may not sound like much, but the average annual deficit is almost 11.4 billion gallons—the yearly needs of 475,000 average Lima residents. (Because annual demand counts only those served by the piped water system, it excludes those 1.5 million Lima residents who are not connected, so the actual shortage is surely higher.)

The researchers calculated that *amunas* in the Rimac River watershed, a larger water source for Lima than the Chillón River, could more than make up that deficit. In their model, they calculated a diversion of about 35 percent of the Rimac watershed's wet-season stream flows into the *amunas*, leaving the rest in the river to support aquatic life. They also made the conservative assumption that half the diverted water would also go to the environment, deep underground or released into the atmosphere via plants' transpiration. Nevertheless, what remained was about twenty-six billion gallons— more than double what Lima needs to make up that 5 percent deficit. And because poor people use much less water than rich people, the surplus could potentially serve a significant portion of people who are currently unconnected too. If the researchers had included the watersheds for the Chillón and Lurin Rivers, which also supply Lima, that twenty-six billion gallons would be even higher.

Another key question the researchers sought to answer was just how much later the water appeared. They discovered that *amuna* water spends between two weeks to eight months underground, with an average delay of forty-five days. Scaling up the use of *amunas* would increase Rimac River flows at the beginning of the dry season by up to 33 percent and could thus postpone the date when water managers must tap Lima's reservoirs.

Studies like this one play an important role in expanding the rollout of Peru's national program. Despite the new laws requiring investment in natural infrastructure, the country's technical, planning, and finance systems are still not really set up for such expenditures. Forest Trends is working with local agencies to overcome these hurdles, and collecting and publishing data is part of that effort. Around the world—and still in Peru—decision makers come from a culture of concrete. They default to gray infrastructure such as dams and levees because these choices are fortified by mountains of scientific papers that precisely quantify their benefits (although often ignoring their hazards), while science proving out natural infrastructure is still young. As Gammie puts it, "Gray infrastructure has a hundred-and-fifty-year head start."

Part of the dilemma is that projects must be built before they can be studied. It's a chicken-and-egg problem, to some extent: Which comes first, the project or the data? Yet when natural infrastructure projects are built around the world, they often don't get enough funding to monitor how they work. That's a barrier to getting international financing, which limits the building of additional projects. That's why research such as the *amunas* study is critical to changing how we manage water. It translates the efficacy of Slow Water projects into the language that decision makers speak.

Ochoa-Tocachi is working on numerous other studies in Peru and elsewhere, contributing to a growing body of research on natural infrastructure projects. Broadly speaking, studies so far demonstrate that these projects that work with nature deliver multiple benefits, often for lower costs than engineered alternatives.

Encouraged by the findings of the *amuna* study, SEDAPAL is investing US$3 million in shoring up twelve *amunas* above Huamantanga, building two more, and restoring the neighboring grassland. It is also investing in restoring and supporting *amunas* in another town. Other Indigenous techniques are inspiring the utility as well. In yet another mountain village, it is looking to restore *andenes*, the system of terraces for farming that captures water and infiltrates it, conserving water for crops and reducing erosion.

Ecosystem Services

Another reason decision makers have been reluctant to support nature-based solutions is more subtle. Our dominant economic system—global capitalism—takes for granted the things nature gives us. Wetlands clean and store water. Trees emit oxygen, exhale water to form rain, and cool the air. Coral reefs and mangroves protect our homes from storms and provide nursery habitat for the wild fish we eat. Organic matter in the soil feeds our food plants. Insects and birds pollinate them. In the economic system, these resources are free, or nearly so.

Harm caused to these resources by industry, such as increased scarcity or pollution, are often not borne by those who profit from this damage but by nature or marginalized people. Economists call these costs "externalities" because they are external to the economic system. In general, oil companies don't pay for dumping their pollution into rivers that Indigenous people rely on for drinking and food. Dam builders don't compensate river animals for threatening their ability to survive, or subsistence fishers for their loss of food. These industries often pay little or nothing for flooding cultural sites, or forcing people from their ancestral villages. (Many places are passing laws to try to remedy these oversights, but industries often apply political pressure to avoid consequences.) This is one major way that the rich get richer and the poor get poorer.

Another critical flaw in mainstream economics that also affects water and the rest of the environment is its goal of eternal growth. Most of our economy depends on natural resources, which are finite, a reality the system does not recognize. When we take them for granted, we don't plan for the day when things run out or ecosystems tip into dysfunction. In fact, standard economics is centered around scarcity: when something becomes rare due to overexploitation, it becomes more valuable, and someone, somewhere, makes a profit. And eventually, industries seek alternatives, starting the cycle again. Global capitalism is not geared toward simply providing enough for everyone, but rather, grabbing as much as possible for the privileged. The goal of extracting something to that extent—a kind of economic extinction—is diametrically opposed to considering the long-term health of the system to which the thing belongs. Global trade tends to exacerbate this phenomenon, because when industries exhaust resources in their own country, or find them cheaper elsewhere, they turn abroad, where the damages are out of sight, giving little incentive to rein in consumption. In contrast, cultures that practice reciprocity, like the *comuneros*, are rooted in a mindset of abundance: you give, trusting that you will also receive what you need.

The logic gap to eternal growth is also evident when demographers argue that the slowing population growth in some countries is a big problem because there won't be enough young workers paying into pension programs to support larger numbers of old people. With this argument, eternal population growth seems to be the goal, as a crutch to uphold our current economic system. It's an assumption that ignores the limits of a finite Earth and all the problems that come with growing populations, from housing, food, and water shortages to ecosystem destruction.

Because the dominant system is constructed to ignore all of this, and because it's geared toward short-term profits rather than long-term sustainability, decision makers may not understand the need to support nature-based solutions. Yet after centuries of taking from and despoiling ecosystems, nature's ability to give freely—with little support or protection—is nearing an end. As ecosystems break down, their services do too. Nowhere is this more apparent than in our interactions with water. "The ecosystems on which life itself is based—our food security, energy sustainability, public health, jobs, cities—are all at risk because of how water is managed today," said World Bank Group President Jim Yong Kim in 2018.

The fields of environmental economics, ecological economics, and steady-state economics attempt to correct these destructive inaccuracies. In her excellent 2017 book *Doughnut Economics*, economist Kate Raworth draws out a model that lives within both planetary boundaries and social foundations for human equality. In a sign that this thinking may be becoming more mainstream, in spring 2021, the United Nations adopted a new framework to include nature's contributions when measuring economic prosperity and human well-being. Going by a catchy name, the System of Environmental-Economic Accounting—Ecosystem Accounting (SEEA EA), the framework aims to include "natural capital" in economic reporting. That includes forests, wetlands, and other ecosystems. This framework would go beyond the long-dominant gross domestic product, or GDP, which ignores the economy's dependence and impacts on nature. In its memo about the new recommendations, the UN said, "More than 34 countries are compiling ecosystem accounts on an experimental basis," and it expected many more to join.

The idea of natural capital is explained in a groundbreaking 1997 study in *Nature* that tried to quantify how much it would cost if we destroyed ecosystem services and people had to try to manufacture replacements. When water is polluted and needs to be cleaned, how much does that cost? What if water isn't available and needs to be produced via desalination? If insects who pollinate crops went extinct and people had to fertilize food by hand, how much would that cost? The price at that time was US$33 trillion a year, a bit more than the global gross domestic product. The researchers updated their study in 2014 to an estimated $125 trillion annually, which was then more than one-and-a-half times the global GDP. With these values in mind, people created the idea of payments for ecosystem services, an attempt to incorporate the value of what nature provides to the economy into the accounting books—to move "externalities" inside the system.

Not everyone working toward sustainability supports this move. They argue that other species and nature itself have a fundamental right to exist separate from whether they benefit humans. It's the same argument made by the lawyers arguing for nature's rights. For that reason, they say that putting a price on the things nature does for humans is unethical. That's because species' and nature's fundamental right to exist makes them priceless, in a sense, and because it makes their use or conservation a market decision—meaning they could be destroyed capriciously. But those who advocate for ecosystem services—including NGOs such as Forest Trends and think tanks such as the Natural Capital Project at Stanford—argue that if you want corporations and governments to value something, you have to translate it into the language they speak: money.

Peru's government-mandated Mechanisms of Reward for Ecosystem Services (MRSE) was influenced in part by earlier projects centered around voluntary payments for ecosystem services. For example, in Ecuador and Colombia, regional water funds with money from corporations or downstream users have financed natural infrastructure projects to support water resources. Although it has a different structure, MRSE, in recognizing the role of natural ecosystems in supplying water, is rooted in the same acknowledgment that ecosystems provide the things humans need.

These ideas might sound radical, but they are not limited to people on the fringe. Consider a letter published in the journal *BioScience* in 2019, titled "World Scientists' Warning of a Climate Emergency." Signed by 11,258 scientists from 153 countries, it called for "bold and drastic transformations regarding economic" policies,

including a "shift from GDP growth and the pursuit of affluence toward sustaining ecosystems and improving human well-being."

*

Still, it can be hard to move from vision to reality. Despite Peru's
national policies embracing investment in ecosystem services, the
government's frequent reshuffling has slowed the rollout.

Back in Lima, in the Pueblo Libre neighborhood, I meet with
Francisco Dumler, a dapper man with a receding hairline in a gray
wool herringbone coat and salmon pink pants. In July 2019, he'd
just been appointed president of the board for Lima's water utility
SEDAPAL. Dumler has worked numerous government jobs during
his career, including in housing, sanitation, and the Autoridad Nacional del Agua (National Water Authority). He also went to school
with Fernando Momiy, creator of the Natural Infrastructure for Water Security concept. I congratulate him on his new position and
ask, given the notable turnover, how long he thinks he will be there.

"Two years," he answers without hesitation.

Blasé about the personnel merry-go-round, he tells me he's determined to build as many projects as possible to maximize his time in
office. "My job is to explain to people who live near the ocean that
we need to be willing to pay those people in the mountains who are
taking care of our water. Because people in Lima still think water
is a matter of opening the faucet and it's magic." He also cites the
concrete culture: decision makers at utilities are schooled in dams,
pipes, and desalination plants. Ecosystems and how they affect water supply are foreign terrain. His brown eyes crinkling as he smiles,
he asks rhetorically, "How many hours do you think I have invested
trying to explain what *amunas* are? Or trying to explain to them how
to restore the land terraces that were originally developed by our
ancient civilizations?"

On the question of which comes first, data or projects, Dumler is
clear: projects. Natural infrastructure interventions are really cheap,
he points out. His philosophy: Just build them, monitor them, and

see what works. To demonstrate how he's hitting the ground running, Dumler flips through WhatsApp, showing me conversations he's having with officials regarding details for specific projects. "I am a true believer in this," he assures me.

Two years later, he's still holding his position and has played a big role in moving projects forward, Gammie tells me. As I write this in mid-2021, forty of the country's fifty water utilities are collecting MRSE funds and have collectively raised more than US$30 million. SUNASS expects them to raise at least $43 million by 2024. Utilities are investing that money in their natural watersheds via more than sixty projects across the country, including the *amunas* and a rare, high-altitude ecosystem that holds water.

Soggy Cushions

Leaving Lima again, this time heading northeast along the Rimac River, I accompany a group of regional water experts organized by Forest Trends to a rare, high-altitude tropical peatland called a *bofedale*, or cushion bog. Unique to alpine valleys in the Andes, *bofedales* are dominated by plants well adapted to tropical mountain conditions, thriving despite intense sun, stiff winds, a short growing season, daily frost, and seasonal snow.

The soils of peatlands, including *bofedales*, have a higher percentage of organic matter than other soils, making them unusually good at holding water. Though peatlands cover just 3 percent of global land area, they store 10 percent of all fresh water (not to mention 30 percent of the world's soil carbon). In the steep landscape of the Andes, *bofedales* slow water runoff, preventing floods and landslides. As glaciers disappear, wetlands play an even more important role in slowing water for supply in the dry season. And because *bofedales* stay green year-round, they are biodiversity hotspots, frequented by birds and mammals, including deer, pumas, Andean fox, pampas cats, and vicuña and guanaco, wild ancestors of domesticated alpacas and llamas.

Though traveling just sixty miles or so, we drive for hours on a winding, slow, single-lane dirt road, up into the clouds. Along the way we have heart-stopping brushes with oncoming traffic and heart-melting encounters with herds of llamas and "burritos"—an affectionate local term for burros that, given my California roots, made me snicker. (Burritos, the tortilla-based food, are not a part of Peruvian cuisine.) Finally, not far past the small Andean village of Carampoma, we reach a spot at nearly fifteen thousand feet elevation where the valley widens, holding a seasonal lake and *bofedale*. But something is terribly wrong. Squares of soil five feet long and a foot deep have been cut out in a checkerboard pattern by peat poachers to sell to plant nurseries in Lima. The remaining patches, freshly stripped to the elements, smell of decay as organic matter oxidizes. The exposed land is hummocked and has a sick, rusty tinge. With our jacket hoods up against the fiercely blowing wind, we stumble across the uneven surface, our footsteps kicking up dust. This peat, laid down over millennia, was destroyed in minutes. Picking up some of the drying duff, a water security specialist with CONDESAN proclaims, "This is a public service announcement: when you are in Lima, do not buy this for your ornamental plants or vegetables. Use compost instead." Her point: it's nonsensical to destroy an ecosystem that supplies water to a desert city to make things grow in said city.

Tragically, the peat thievery happened while Forest Trends and other local NGOs were trying to convince SEDAPAL to invest in this area's conservation. Gammie shows me pictures on her phone taken in early 2016. In them, the valley is carpeted in lush green, an idyll like the mythical Shangri-la or the alpine meadow where Maria kicks up her heels in *The Sound of Music*. The first time Gammie saw it in 2013, she reminisces wistfully, it was raining up here, after a long, dry drive. "I thought, 'It's a sacred place. It's where this river starts. And it's one of the only places where it actually rains in this watershed.'"

But in March 2021, their years of effort paid off. Officials from SEDAPAL and several national ministers traveled up the long road from Lima to Carampoma for a ceremony to kick off the water utility's US$850,000 investment to restore the devastated area and to protect the healthy *bofedales* that remain. The project will work with

FIGURE 6.3 People stealing peat for the plant nursery trade have devastated this important ecosystem. Water detectives are hoping that if they restore indigenous plants and reconnect paths for water to the area, this *bofedale* will recover.
Photo © Erica Gies

the community to move grazing away from affected areas and to introduce surveillance of the *bofedales*.

Peru has laws to protect wetlands, but enforcement jurisdiction is murky. To clarify the situation, Forest Trends is meeting with authorities and developing a manual for the community. The guidelines will instruct local people about what to do (such as take photos and GPS coordinates) and which authorities to notify. To restore the damaged wetlands, people will reintroduce plants harvested carefully from a nearby site and ensure water flow to support them. Scientists don't know how long it will take to restore the peat, but Forest Trends hopes that nature can start to heal quickly with a little help.

Fortunately, these highlands still have some healthy wetlands, and I finally get to see them. Leaving the desecrated peat, we round a hill and another alpine valley lies before us. Low-growing, spongy-firm cushions of green are pocked with wee alpine flowers. Interspersed among one of the dominant plants, a species of *Distichia*, are tiny pools of water. This terrain fronts an aqua blue glacial lake and, behind, a bare mountain whose glacier has melted within Carampoma villagers' lifetimes. Amid the severe dun and rust of the mountains, the green and aqua are beacons of life.

In all these projects, benefits to the local community are crucial so people are motivated to keep up land and water management practices that ultimately benefit the wider watershed.

Despite the slow start to putting money collected from water bills toward nature-based projects, the fact that SEDAPAL is conserving these wetlands is a sign that attitudes are changing. The legacy approach from gray infrastructure is often to build *something*, to have a physical intervention to point to as an accomplishment, rather than to protect a natural ecosystem. An example of this instinct is a misguided earlier project by Cusco's water utility that was a precursor to the natural infrastructure program.

In a bid to shore up the city's water supply, the utility aimed to capture water running off mountains due to overgrazing by infiltrating more rainwater underground. In fact, the natural high Andean ecosystem above Cusco is made for retaining water, from the soil to the iconic tufts of golden bunchgrass called *ichu* (Quechua for "straw"). The grass helps hold onto precipitation and move it underground, explains Luis Acosta, the SUNASS supervisor: "*Ichu* is a highly efficient plant that doesn't consume a lot of water because it has spiral leaves that prevent evapotranspiration." And its deep

FIGURE 6.4 The *bofedale*, or cushion bog, is a rare ecosystem of tropical mountains. These wetlands slow water, extending its availability into the dry season, preventing land-slides and floods, storing outsized quantities of carbon, and providing habitat for many birds and animals.
Photo © Erica Gies

roots help capture rainfall by creating a path through the ground for water to move.

But rather than restoring that natural landscape, project managers for Cusco's water utility wanted to build something. They sliced dozens of close-set horizontal trenches into the flanks of mountains, giving them the appearance of having been combed by a giant Zen gardener. But the cuts they made further degraded the *ichu*, so within a few years, the trenches eroded badly. When I visited, the current project manager was replanting *ichu* along the downslope edges of the trenches to hold the soil. Ochoa-Tocachi coauthored a 2020 analysis of multiple international studies on whether infiltration channels such as these actually move water into the soil. It's not clear that they do, the researchers concluded, and suggested that natural land cover may provide greater hydrological benefits.

Yet even projects focused on using nature can be misguided. "Sometimes we make decisions that are just passionate," Ochoa-Tocachi explains. "We say, 'Oh we want to conserve nature. So we have to plant trees here.'" But planting trees in a grassland—which has been tried in Peru's *puna* and in Ecuador's similar *páramo* ecosystem—can reduce water availability because the trees consume more water than the native grasses. Outside Cusco, I saw *Eucalyptus* uprooted as another earlier program was reversed when people saw the quantity of water these trees slurped in this delicate ecosystem.

Solutions that are designed for a specific place—in harmony with the native water, flora, soil, geology, topography—can better avoid these pitfalls. Missteps like the trenches or the *Eucalyptus* plantings can sow doubt in decision makers and the public about natural infrastructure more broadly. That's why scientific studies and valuing Indigenous knowledge are so important in guiding projects.

"We don't have all the information we'd love to have today to make the best possible decisions," Gammie admits. "But we can make good decisions." Fortunately, SUNASS, the national water regulator, is requiring regional utilities to monitor and report on all Mechanisms of Reward for Ecosystem Services projects. That will allow people here to learn and improve as they go. Over time, Peru's body of research will make a significant contribution to this growing field.

That will be important as water managers globally strive to find new ways to increase the flexibility and resiliency of their own communities' water supplies. Previously many managers made twenty-year plans based on relatively stable water patterns and long-term investments in dams and reservoirs, Gammie explains. But making those same investments today can lock managers into a solution that fails. Rain to fill those reservoirs may not come, or a dam built to specs based on historic weather may not be able to hold the runoff from the giant storms we're seeing today.

Nature-based water solutions, on the other hand, like solar and wind in the energy field, are smaller and distributed across the landscape. They can be built more quickly than, say, a large dam, which can take ten years or more. Nature-based solutions can be scaled up as needed, or more easily adapted as necessary after people see how well they work. In Peru, the *amunas* are modular and adjustable, says Ochoa-Tocachi, and they cost about one-tenth that of gray infrastructure.

Expanding Slow Water solutions across watersheds, as Peru is doing, has a learning curve nearly as steep as the Andes. But the country's shift to natural water infrastructure is not a drill. With its ever-growing water insecurity and the seriousness of the climate situation, Gammie concludes, "we don't have much time—maybe five to ten years to make this move."

· 7 ·

Let Floodplains Be Floodplains

ANTIDOTE TO THE INDUSTRIAL ERA

Water people have a joke: There are two kinds of levees—those that have failed, and those that are going to fail. So how did we become so reliant upon what is fundamentally unreliable—and now even more so with climate change? How did our dominant culture move from close observation of what water wants and working within that—as with the early Tamils in South India and the Huari in the Peruvian Andes—to the hubris of trying to control it? The answer goes beyond our relationship with water to our relationship with nature more broadly. At its root is an expedient cultural separation of human animals from nature that has allowed us to exploit it without conscience.

In Judeo-Christian thought, the idea that nature exists for the benefit of humans stems from the Bible's book of Genesis. Humans shall "have dominion" over all the other animals. Humans shall be fruitful and multiply, "to fill the earth and subdue it." The Judeo-Christian tradition is relevant because it is the basis for much of the dominant culture exported around the world, first via European and American colonialism, then via international lending requirements. (Although some Christians today argue that to "have dominion over" calls for stewardship. Loving god means taking care of the world he made. They call this Creation Care.)

During the Middle Ages, Christian scholars made this idea more explicit with the Great Chain of Being, a hierarchical organization of all life, with god at the top, followed by angels, kings, other hu-

mans, mammals, birds, fish, plants, and minerals. They were inspired
by Aristotle, whose ancient Greek biology text, *History of Animals*,
ranked vertebrate animals first, then invertebrates, then plants. In
the seventeenth century, French philosopher René Descartes argued that humans were the only rational beings and were therefore
separate from and superior to nature and other animals, who were
mindless automatons, merely reacting to stimuli—an idea that has
since been disproven by scientists again and again.

The popularity of human supremacy just ahead of the Industrial
Revolution was a convenient way to justify the exploitation of nature for human gain. This sentiment is deeply ingrained in our modern economic system. We see it when people argue that conserving
a natural area or a species comes at too high a price when it costs
human jobs and corporate profits. The conservationist Aldo Leopold decried this one-way-street mentality, saying, "The [human]-
land relation is still strictly economic, entailing privileges but not
obligations."

With colonization, seeing nature—as well as Indigenous people—as "other" became even more entrenched. Colonizers who
were unfamiliar with the far-flung places they were encroaching
found new-to-them, terrifying threats to human life: locusts who
denuded crops; panthers who stalked them home through forests;
long, harsh winters that brought disease and starvation; swamps
that consumed wagons and bred disease-bearing mosquitoes. The
feeling that nature was callous or even out to get people led to an
us-against-them mentality toward the natural world. It was war,
and colonizing humans often lost. Victories were to be celebrated,
turned into stories, then blueprints.

Water's relationship with people in the Industrial Era took a dramatic turn from the one it had with earlier humans, not just in philosophy but in scale. The balance seemed to tilt in humans' favor
when mass production, technology, and fossil fuels amplified our
power. But without a compensatory change in viewpoint to temper
our impact, what first allowed us to win some of these battles with
nature eventually unleashed a scale of destruction with which we are
only now coming to grips.

In pursuit of control over water, we have indulged in hyper-engineering—building levees, channelizing and damming rivers, and burying creeks and wetlands. As our attempts to tame water have grown in scope, we've raised the stakes for failure, both in agricultural areas, such as the US Midwest and parts of the Netherlands, and in rapidly expanding cities, such as in China. When water takes what it wants by force, the results aren't pretty. Around the world, the number of people affected by floods is expected to double between 2010 and 2030, according to World Resources Institute's Aqueduct tool. Along rivers, numbers affected will rise from 65 million to 132 million annually, and costs from $157 billion to $535 billion. On the coast, people affected will rise from 7 million to 15 million, with costs rising from $17 billion to $177 billion.

In response to disasters, places like the Netherlands and China are starting to unwind the industrial approach to give water space on a meaningful scale. In so doing, they are setting a course correction for the assorted ills we've brought upon ourselves during the Industrial Era.

We have lost much in seeking control: soil and land retention, natural coastal protections, water cleaning services, carbon storage and regulation, knowledge of how to live within the limits of our water resources, habitat for many plants and critters, more livable habitat for ourselves. But perhaps the most in-our-face consequences are the abrupt failures: massive flooding of our farms and cities.

Control, as it turns out, is illusory. Have you heard the one about the levees?

The Dire Straits of Levees

Levees failed spectacularly in the central United States during winter, spring, summer, and into fall 2019. Challenged by extreme rain and snow events (harbingers of climate change), dozens of levees broke and hundreds of miles of river barriers were overwhelmed. Along the Mississippi, Missouri, and Arkansas Rivers—the flowing heart of the country—parts of nineteen states were flooded, some for up to seven months. All told, the disaster cost $20 billion in losses and twenty million acres of farmland to go unplanted. In southern Louisiana, the Mississippi River stayed above flood stage for 226 days. It was the most extensive and prolonged flooding in the region's history. The massive disaster was a harsh reckoning for people who had responded to previous floods by building levees, first by hand, later by bulldozer. Today the Mississippi alone has 3,500 miles of levees averaging twenty-five feet high. But to water detectives, the extreme flooding was unsurprising.

Building a levee on one bank pushes high waters onto the opposite bank, launching an engineering arms race. People on the other side build levees too, and the water, confined to an unnaturally narrow channel, unable to spread and slow upon its floodplains and wetlands, rises up and flows faster. The longer the length of levees, the greater the constriction, the higher and faster the water, the worse the flooding becomes downstream and at every place where a levee breaks. A narrowed river can even cause flooding upstream by creating a bottleneck.

The US General Accounting Office acknowledged these hydrological truths with regard to both the Missouri and Mississippi Rivers in 1995: "That levees increase flood levels is subject to little disagreement." Studies noted that flood levels near St. Louis had increased up to thirteen feet over the twentieth century.

Unfortunately, these effects are generally recognized only in hindsight because proposed levees are evaluated individually without considering the cumulative effects of many projects along the length of a river. Even the Army Corps of Engineers, a big proponent and builder of levees, has recognized this reality, acknowledging that levee permits, granted on an individual basis, result in a "death by a thousand blows" through the "incremental loss of floodplain land to development."

Ironically, levees also multiply flood damage by giving people a false sense of security. By preventing small floods, a levee reduces the frequency of flooding and makes an area appear dry. That encourages development and draws new people into the floodplain, where their new homes and businesses are vulnerable to catastrophic floods when levees overtop or fail.

This phenomenon is called the "levee effect," and was first described in 1945 by the geographer Gilbert White, who's known in water circles as the "father of floodplain management." Floods are necessary for many natural processes, as we saw in chapter 3. Humans view flooding as a problem only when people are affected, but we greatly increase the risk of impacts by building in floodplains. In his doctoral dissertation, White argues against engineering solutions to floods. Instead he advocates for modifying human behavior (such as by not building in floodplains) to reduce potential for harm. "Floods are acts of god," he writes, "but flood losses are largely acts of man."

The levee effect has been getting more attention from researchers in recent years, particularly with the recent emergence of the field of sociohydrology that studies how human behavior interacts with water and infrastructure. For example, researchers have documented how a community's experience of a major flood prevents further development in the floodplain for a while, but that communal memory decays, often within a decade. A notable case is the historic flood in 1993 that left the Mississippi River above flood stage for 147 days, lapping at the foot of the iconic Gateway Arch in St. Louis. Yet, twelve years later, the city had built twenty-eight thousand new homes and more than ten square miles of commercial and industrial development on land that had been inundated.

The story of how we came to levee up one of the world's most dynamic river systems is grippingly documented in the 1998 book *Rising Tide* by historian John M. Barry. Nineteenth-century engineers and other people fought over how to tame the Mississippi River and its tributaries. Renowned civil engineer Charles Ellet argued for a multipronged approach to mitigate flooding—with levees, yes, but also bypass channels and outlets that people could open during high flows, and wetlands to absorb rainwater. His ideas were competing with those of A. A. Humphreys of the US Army Corps of Engineers, who argued that the country should take a "levees only" approach. Proponents argued that closing off natural river outlets would constrict the flow, forcing the river to scour out its channel, making it deeper, both for shipping and to convey big floods downstream safely. The corps prevailed, and "levees only" dominated for decades. But engineers also added "river-training structures"—such as wing dams perpendicular to river flow—to confine water in dry summer months to keep it deep enough for shipping. These structures act as barriers when high flows come and actually raise flood levels many feet.

Broadly speaking, today's midwestern culture remains resistant to giving water space because of a deeply ingrained tradition of control. Journalist Sharon Levy explains why in an article for *Undark Magazine*. Part of what's now the Corn Belt was once the Great Black Swamp, "a wild expanse of wet forest and marsh stretching across a million acres," she writes. European colonizers felled giant sycamores and oaks and dug "miles of drainage trenches to slowly bleed the water away from the muck." Innovation eventually led to "underdrainage": digging channels beneath a field and lining them with pipe to funnel away excess water. With these techniques, many Corn Belt states—including Iowa, Illinois, Indiana, Ohio, and Missouri—banished 85 to 90 percent of their native wetlands. The blood, sweat, and tears that generations of people invested in drying out the land means today's residents are resistant to restoring wetlands, Levy writes. "They've passed down a deep and abiding loathing of wetlands. They are considered a menace, a threat, a thing to be overcome." An Ohio state law even forbids anything "that slows the flow of runoff through those miles of constructed drainage ditches"— including wetland restoration.

In the early 2010s, the Omaha and Kansas City districts of the Army Corps made some small levee setbacks along the Missouri to give the river a bit of access to its floodplains. For their initiative, they were sued by 372 property owners who claimed that, with the predictable flooding the setbacks allowed, these actions amounted to a property taking for which they had not been compensated. The landowners won an estimated $300 million in damages. What tends to win public support in the Midwest is hard engineering, what one expert calls "military hydrological complex projects."

But of course, not all Midwesterners think this way. Olivia Dorothy is director for the Upper Mississippi River Basin for the NGO American Rivers. She lives in East Moline, Illinois, which flooded in 2019 when a levee broke. What was unique about that year was the duration of the flood, she tells me by phone. (Although with climate change bringing increased rain to the Midwest, that could well be repeated.) The levee system was designed to withstand brief flood events: "a week or month at most. When you have water on levees for so long, they get supersaturated—they fail." She explains why: "Rivers don't just exist on the surface. Even if you change out sand levees for clay, or even concrete, the water can move under it because it's connected to the groundwater and aquifer."

A few places along the 2019 flood course had previously taken a Slow Water approach. Davenport, Iowa, which sits along the Mississippi River, decided in the early 2000s to restore Nahant Marsh, a 305-acre wetland right near the urban area, and to use riverfront parks to accommodate high flows. In 2019, Nahant Marsh absorbed as much as a trillion gallons of floodwaters, protecting much of Davenport. Eventually part of the downtown flooded, but the damage was much less severe than it was elsewhere. Marsh managers bought an additional thirty-nine acres of farmland in 2018 between Nahant Marsh and the Mississippi River, restoring it to wetlands and prairie. Now they are looking for more. The city is embracing the wisdom of a local professor of civil and environment engineering to "let floodplains be floodplains."

Small-scale relinquishment of control to give water some space to do what it wants is becoming more common both in agricultural areas and in cities. However, most initiatives so far have been relatively limited, like Nahant Marsh. When faced with floods like 2019's, the problem is evident: the space we are ceding is too meager. To protect people and property when flows are as high as they were in 2019, a river needs access to its floodplains along more of its length. If additional cities along the Mississippi joined Davenport in removing some levees and making riverside land available for periodic flooding, their cumulative efforts would be better able to slow the rush of water, each absorbing part of the flood, thereby reducing risk for everyone. Restoring wetlands on just 1.5 percent of a landscape can reduce the height and volume of flood peaks by as much as 29 percent.

Stepping Back from Industrialization: Room for River

Cultural change is hard, but not impossible. The Dutch were among the first to levee up at industrial scale—and then to recognize that the jig was up. While still industrially minded, the Dutch have adjusted their basic philosophy. Henk Ovink, a Dutch landscape architect and evangelist for natural flood solutions, is the Netherlands' special envoy for international water affairs (and worked with local water managers in Chennai). He says the key is to "try to live with water instead of fight it."

The Dutch had danced with the sea and four river deltas for nearly a millennium, building what they call dikes (levees) to enclose rivers, block the sea, and thus protect low-lying land. But in 1953, a strong storm blew out their dikes—a disaster terrifyingly documented in the 2011 book *Blue Revolution* by journalist Cynthia Barnett. The storm hit during a high spring tide and generated waves up to thirty-two feet tall. Water rushed through the breaks like Niagara Falls. Ultimately, the storm and its floods killed 1,835 people and fifty thousand cattle and destroyed or damaged forty-three thousand homes and 3,300 farms, Barnett writes.

The traumatized Dutch responded by building megadikes, a project that would take four decades to complete. As estuaries were closed off from the sea, water behind the dikes grew stagnant. Ecosystems died, birds and fish disappeared, pollution accumulated, algae bloomed. And despite that heavy price, the megadikes didn't solve flooding. With water's natural escape to the sea closed off and climate change and development causing bigger river floods, in 1995, water rose precipitously in the Rijn (Rhine), Maas (Meuse) and Waal Rivers, forcing 250,000 people to evacuate. Although the waters eventually receded without flooding towns, "it raised the awareness of how vulnerable our riverine system was," Ovink tells me.

After 1995, the government decided it needed a different approach. The program, called Room for the River, which Ovink helped formulate, increases the water capacity of river deltas by removing or lowering dikes to reconnect some stretches of river with their floodplains.

Farmers were asked to cede land that had been in their families for generations. Communities across the region came together to discuss the problem with experts and officials and ultimately agreed on solutions. That approach worked because the Dutch have been collaborating with each other to manage water since the 1100s, Ovink says. They've had to because, if a farmer built levees around his house, the water would then inundate neighbors. "Our democracy started as a water democracy with regional water authorities, even before we were a kingdom."

Still, it wasn't easy or quick. Democracy takes time. The government didn't tell people they had to leave their houses. Rather, it said, "'these are the risks, here are some possible solutions. . . . We have to make room for the water,'" recalls Ovink. Some farmers were persuaded to give up their land; others agreed to let their land flood when necessary, after evacuating their cattle to higher ground.

Room for the River is not a fully nature-based system. Although it uses floodplains, it also relies on levees. Some Room for the River projects have another levee behind the first as a second line of defense. The difference is that if the levees fail, the consequences will be less severe than in a traditional confined river system.

Expanding on the premise of reducing the consequences of failure, the Netherlands updated its Water Law in 2017 with something called the risk norm. The idea is to shift from trying to prevent floods to managing risk from floods. The risk norm evaluates a place based on the probability of flooding and the scale of consequences. Basically, risk = probability × consequence. In a place with expensive infrastructure, people take action to reduce the probability of flooding to minimize risk. The consequences of this policy (although not made explicit in the law) would mean giving water somewhere else to go, thereby letting the probability of flooding increase in places with lower cost impacts.

Sponge City Revolution

Across the Eurasian continent in Beijing, China's leaders have adopted a related credo to make room for water in cities. They too were driven to action by disaster. Between 2011 and 2014, 62 percent of China's cities flooded, costing $100 billion in economic losses. That percentage may sound shocking, but China is not unique. In the United States, for example, in a 2018 survey of municipal stormwater and flood managers, 83 percent reported flooding in their areas.

Urban flooding has become particularly acute in recent decades, as the world undergoes a mass migration from rural areas to cities. In 1960, fewer than 34 percent of people around the world lived in urban areas. It was 55 percent by 2018, according to the United Nations, and is likely to hit 68 percent by 2050. And because the world's population has more than doubled since 1960, the number of people living in urban areas has jumped from one billion then to more than four billion today.

To accommodate the new urbanites, the land area covered by cities and their impermeable pavement has doubled worldwide since 1992. Greater New York occupies 4,669 square miles; Tokyo–Yokohama, 3,178; Chicago, 2,705; Moscow, 2,270; Houston, 1,904; Beijing, 1,611; Johannesburg–Pretoria, 1,560. Researchers from Johns Hopkins University calculated how impervious surfaces increase flooding: every time a city increases coverage of absorbent soil with roads, sidewalks, or parking lots by one percentage point, runoff boosts the annual flood magnitude in nearby waterways by 3.3 percent. The many cities that were built in part on filled-in wetlands—from New Orleans to New York, London to Paris, Houston to Venice and Shanghai—are particularly prone to flooding because water still wants to follow its innate path.

Nowhere has urbanization happened more rapidly than in China, where a mass exodus from the countryside over the last forty years has seen the number of urban dwellers boom, from 20 percent of the population in 1980 to 64 percent in 2020. To house and employ all these people, cities sprawled and new ones were built from scratch. Builders paved floodplains and farmland, felled forests, and channelized rivers, leaving stormwater that once filtered into the ground nowhere to go but up and over levees. Then, one notable flood struck the national government where it lives. Poked in the nose, the dragon changed course.

*

Up to eighteen inches of rain chucked down on greater Beijing on July 21, 2012, flooding roads three feet deep and filling underpasses. Landscape architect Yu Kongjian barely made it home from work. "I was lucky," he says. "I saw many people abandon their cars." As the deluge fell, the city plunged into turmoil. Beijing's largest storm in more than sixty years killed seventy-nine people, most of them drowned in their vehicles, electrocuted, or crushed under collapsed buildings. The damages stretched across 5,400 square miles, costing nearly $2 billion.

Yu, cofounder of the internationally acclaimed landscape architecture firm Turenscape, was not just saddened; he was frustrated. He'd warned the government years earlier that disaster was coming. He had led a research team in mapping what he calls the city's "ecological security pattern," showing the government which parcels of land were at high risk for flooding and urging it to block development and instead to use them to absorb stormwater. They ignored his recommendations. "The 2012 flood gave us the lesson that the ecological security pattern is a life-and-death issue," Yu told me when I met him in Beijing in 2018.

This urban sprawl is also exacerbating water scarcity in China, especially in the north and west. In some of China's dense cities, due to rain running off buildings, streets, and parking lots, only around 20 percent of precipitation actually infiltrates into soil. Instead, as in so many other cities around the world, drains and pipes funnel it away—lunacy, Yu thinks, in a place with water shortages. Like other cities in China's north, Beijing is pretty dry outside the summer monsoon season. For decades it has pumped groundwater to supply its growing population and consumption rates. The city is lowering the water table about three feet a year, causing the ground to sink as well. This phenomenon is also happening elsewhere, such as Mexico City and California's San Joaquin Valley, and it can permanently reduce the capacity of the aquifer as space between gravel and sand becomes compacted.

But now Yu—a slim, intense man with shrewd eyes and just a bit of gray hair at the temples—is leading the way as China reengineers old cities and designs new ones to accept rather than fight natural water flows. His landscape architecture projects incorporate Slow Water principles to lessen floods, save water for dry spells, and reduce water pollution.

The 2012 Beijing disaster was a turning point. A month later, a Turenscape stormwater project in Harbin, a city about eight hundred miles northwest of Beijing, won a top US design prize. China Central Television broadcast a high-profile interview with Yu. He says a government minister told him afterward that President Xi Jinping had seen it. Less than a year later, the president stood before China's national urbanization conference and announced his Sponge City initiative, boosting the idea from fringe concept to national mission. (A sponge absorbs water, then releases it slowly.) It's part of Xi's Ecological Civilization, his agenda to clean up the pollution, flooding hazards, and associated costs caused by his predecessors' Industrial Civilization. With its centralized government, China built industry and its economy at a blistering pace. Similarly, now it is pursuing Sponge Cities on a scale difficult for most countries to even consider.

In 2015, the government began demonstration projects in sixteen cities, adding fourteen more in 2016. Each project covered at least five square miles, although some were much larger. Sponge City objectives include reducing urban flooding, retaining water for future use, cleaning up pollution, and improving natural ecosystems. The goal by 2020 was for each project to retain 70 percent of the average annual rainfall on site both to help prevent flooding and to infiltrate water underground for the dry season.

One challenge is financing. The Ministry of Housing and Rural–Urban Development invested an average of 500 million RMB (about US$72 million) per year for three consecutive years in each of the first sixteen pilot cities. At the same time, it estimated that the total cost of the projects for the sixteen cities would run 86.5 billion RMB (US$13.5 billion). So the amount it invested was about 28 percent of the cost. The central government appealed to local governments and to private investors to pony up the rest. But Yu tells me that Sponge City projects are a hard sell to the private sector because the payoff is the public good, not profits for investors.

Despite these challenges, the pilot cities met their stated goals— according to China, says Chris Zevenbergen, an expert in urban flood-risk management at the IHE Delft Institute for Water Education in the Netherlands and a visiting professor at China's Southeast University. "It is, of course, a success story." The state-controlled broadcaster China Central Television reported in November 2020 that the thirty pilot projects had been completed and were significantly preventing and mitigating urban disasters, increasing environmental benefits to waterways, and reducing water pollution. While Chinese government reports are best viewed with some degree of skepticism, signs point to the country's ongoing commitment to Sponge Cities. Difficulty accessing data about the pilots may be down to a lack of monitoring data and evaluation standards, Zevenbergen tells me by phone. But the central government published evaluation standards in 2019, so more data may be coming. And in 2020 it urged all pilot cities to submit self-evaluation reports on their Sponge City construction.

China has already expanded the national program beyond the pilot cities. CCTV further reported that between 2016 and 2020, the Sponge City concept was implemented in ninety provincial-level cities and was included in the master plans for 538 cities. The goal was for 20 percent of cities with populations of more than one million people to capture 70 percent rain on site by 2020. The new target is for 80 percent of such cities—of which China has 125—to meet the goal by 2030.

Landscape-design projects that aim to slow water are popping up in cities everywhere but are typically quite small, shoehorned between buildings or in narrow corridors along streets. In Houston, developers of new apartment complexes dig bioswales as an alternative to concrete stormwater detention cisterns. San Francisco jackhammers bits of sidewalk or roadway medians for plantings that can absorb a little rain. Even the scale of China's recent projects is likely insufficient. During heavy rains in 2021, one pilot city, Zhengzhou, still suffered significant flooding and deaths. Absorbing rainfall across five square miles of a city that spans thousands wasn't enough to avert disaster.

But Yu and other urban water detectives are looking to manage water to a grander extent, seeking out connected routes for water to slow and flow across an entire watershed, which often extends beyond jurisdictional boundaries. China's pilot projects have shown the challenges of coordinating across these administrative lines. But such issues must be worked out, because solving a city's flooding problems requires coordination with communities and landowners upstream, just as Lima, Peru, is looking to mountain towns to address water scarcity. To really make a dent in city flooding, urban designers would ideally aim to absorb water where it falls across the watershed, reducing stormwater runoff at every rooftop and at every farm field upstream. Yu is keen, dreaming beyond Sponge Cities to Sponge Land. "This is a philosophy for taking care of the continental landscape," he tells me. "It's time to expand the scale."

Inspired in part by American landscape designer Warren Manning, who created a "National Plan for America" a century ago, Yu is working on a landscape master plan for all of China. He has a series of maps that document the country's elevation, watersheds, flood paths, biodiversity, desertification, ecological security, soil erosion, and cultural heritage. Niall Kirkwood, a professor of landscape architecture and technology at the Harvard Graduate School of Design, who has known Yu for many years, tells me via phone, "That is an incredible vision. No one thinks at that scale and with that political savviness."

Yu tracks China's landscape changes as urbanization spreads, as estuaries and deltas silt up, as water starts to move differently across landscapes and cityscapes. He can isolate priority areas where projects will have the biggest impact. Kirkwood compared this to applying acupuncture to the human body. Yu "understands that doing a piece of work in one area will have an effect in another area," Kirk-wood explains. He is "thinking much more holistically" than most landscape architects.

Mr. Sponge City

On a spring day with a "very high" air pollution rating, I visit Yu at Turenscape's headquarters in Beijing's northwest Haidian District. In his plant-filled office, he tells me that he traces his passion to re-pair humans' relationship with water back to the agricultural com-mune where he grew up, in Zhejiang Province southwest of Shang-hai. There he observed the Chinese "peasant wisdom" for managing water, practiced for thousands of years. Farmers maintained little ponds and berms to help rainfall infiltrate the ground, storing it for a dry day. The seasonal creek next to his village swelled and retreated with the seasons. "For me, flood is a time of excitement because the fish come to the field, the fish come to the pond." He saw that flood-ing need not be the enemy. "If you have wise ways to deal with flood, water can also be friendly."

Since starting Turenscape in 1998 with his wife and a friend, Yu has built the award-winning company into a landscape architecture empire, with six hundred employees in three offices. The company has more than 640 projects built or underway in 250 Chinese cities and ten other countries. Yu is also founder and dean of the College of Architecture and Landscape at Peking University and has taught periodically at Harvard.

For years while Yu was building his firm's portfolio, many Chi-nese derided his farm-based ideas as backward. Some even called him an American spy—a nod to his PhD from Harvard's Gradu-ate School of Design and his opposition to dams, those symbols of power and progress in modern China. But since President Xi began championing Sponge Cities, sentiment has begun to shift. Various groups in China are building green infrastructure projects, often in partnership with Americans, Australians, and Europeans. Yu's influ-ence has been growing in parallel.

He lectures regularly at the Ministry of Housing and Urban–Rural

Development, and his 2003 book—*Letters to the Leaders of China: Kongjian Yu and the Future of the Chinese City*—is in its thirteenth printing. He's been asked to consult and lecture in other countries, such as Mexico, which is hoping that he can help solve Mexico City's water problems, which are similar to Beijing's.

While Yu retains his farmer's values, he is a man of modern China. He bought and renovated a building in one of Beijing's few remaining historic *hutong* neighborhoods, turning it into a private club for fellow Harvard grads, Beijing politicians, and other power brokers. Later that day, he escorts Peter and me through engraved metal doors and across the courtyard, where the traditional stone floor has been replaced with thick glass. Inside he ushers us downstairs to a massive table underneath that transparent floor. As we sit in ornate, carved chairs sipping bright-green cucumber juice, I look up and see the moon above. Finance ministers are also visiting the club this evening, so Yu rotates between our tables at intervals. After dinner, his driver drops us at our hotel, and he gets out to walk home, about twenty minutes away, part of his daily constitutional.

Making Maps as Digital Twins

When planning a project, Yu and other urban designers start by trying to figure out what water did before a city spread, and what it does now within its current confines. In a large white room at Turenscape's offices, young men and women sit at desks separated by copious green plants, focused intently on that question. They are building geographic information system (GIS) models of how water behaves in today's city based on cultural, physical, biological, and hydrological patterns across the landscape, a kind of computational geography.

Like many of the water detectives I met, they use spatial mapping software from Environmental Systems Research Institute. ESRI was founded by a low-key, humble guy named Jack Dangermond and his wife Laura. Jack Dangermond grew up in Southern California in the mid-twentieth century. His parents owned a plant nursery and the whole family spent a lot of time in nature. Like Yu, Dangermond also studied landscape architecture at Harvard (though a bit earlier), where he was introduced to making maps on computers. He realized there was a database underneath the maps, and this inspired him to build databases that would allow him to model natural processes and human behavior. ESRI can map watersheds from mountains to ocean, modeling floods, plant succession, infrastructure, and much more to help us move beyond single-issue problem solving. The tool allows us to get our minds around complex systems and their inter-related challenges, such as how to prevent biodiversity loss, build smarter cities, and reduce resource waste. A specialized add-on called arcHydro includes data from hydrologists to better predict rainfall and flooding.

The first thing planners plot is topography—the highs and lows of the landscape—a primary factor in how water flows. While this data can come from satellites, their resolution is typically insuffi-cient for cities, says Ryan Perkl, the head of green infrastructure at ESRI. More refined data can be gathered by airplanes with Lidar (light detection and ranging) sensors, which use lasers to survey urban topography under buildings. City maps can then be overlaid, showing transportation corridors, existing parks, people's yards, and industrial buildings with giant roofs.

Models include soil type, which, as we've seen, can dramatically affect how water drains. They also include vegetation, because that affects how much water soaks in, runs off, or vaporizes from plants into the air, explains Perkl. And soil's pH can affect which plants will thrive or die in a restored area. But getting good soil data in a city can be tricky. In the United States, for example, soil inventories come from the US Department of Agriculture, which "has no interest in urban landscapes," he says. Also, builders often move soil from one place to another, changing its natural composition while flattening the ground for foundations. To get reliable soil data, engineering firms typically drill a hole and take a core sample. If soil at a project site doesn't infiltrate water at the desired rate, the people designing green infrastructure often excavate clay and mix in sand.

Turenscape also models historic and ecological data as well as information on population, economy, and transportation. With this information, Slow Water practitioners can better understand how a variable affects the way water behaves. When their landscape maps are complete, they send test floods through the digital "twin" they've created. These experiments allow them to identify pinch points where water is constrained and will flood first. Then they experiment with a topography adjustment or an addition of a wetland or pond to see how it impacts stormwater behavior.

Of course, to build an accurate model, you need accurate data. In many rapidly growing places like China, this data is currently lacking, Perkl tells me, but China is launching national initiatives to gather them. Zevenbergen is more concerned about access. In his work there, he's found that "data sharing is a big issue . . . constraining collaboration."

A River Reconstructed

A week after meeting Yu, I visit one of Turenscape's projects in progress, Yongxing River Park, located in Daxing, a far-flung exurb of Beijing. "Before" satellite pictures from three years prior show open land surrounding the river, which was straightened and confined by steep concrete walls. "Now" pictures are chock-a-block with buildings around a more generous, meandering path for water. Showing me the park are two of Yu's employees, Geng Ran and Zhang Mengyue. Geng and I bond over our cats, sharing pictures and giggling that hers has an American name, Michael Jordan, while mine has a Chinese name, Chairman Mao Mao.

The Chinese government understands that development reduces rain infiltration, Zhang tells me as we walk along the river, so it invited Turenscape to design a park that would enlarge the riverbed to hold more water. "We say you can't stop the weight of the river," Geng adds, "so that's why we enhance this river."

The project was nearly complete when I saw it in April 2018. About two-and-a-half miles long and perhaps two city blocks wide, the park follows the river. Workers removed concrete along the river channel and excavated soil to widen the riverbed. That dirt was then molded into a large berm running down the center, creating two channels. The river flows on one side; the other channel has big holes of varying depths that act as filtration pools and direct the water flow. During the dry season, the filtration side is filled with partially cleaned effluent from a sewage treatment plant. Wetland plants in the pools slow the water, further cleaning it and allowing some of it to filter into aquifers. During the monsoon, that channel is reserved for floodwaters, and the effluent is treated industrially.

We walk a slim concrete path atop the central berm. Many of Turenscape's designs feature walkways such as this, soaring above wetlands, so people can enter the landscape even when it's wet. This park also has round cement shelters where people can linger and enjoy the landscape. Cut with abstract geometric openings, they impart a vaguely Flintstones flavor. The broader riverbanks, newly freed from concrete, are dotted with thousands of small sedges planted in closely set rows to hold the earth, like a pointillist-rendered landscape. As we walk the line, we pass young willow trees scaffolded together with sticks for strength while they grow. Willows, that native streamside plant that beavers love, have roots that reach for the air, like cypress and mangroves, allowing them to survive extensive periods of flooding. Elsewhere, reeds, small bushy willows, dwarf lilyturf, and other native plants stabilize the soil. Existing large trees, including elms and poplars, were retained.

During big rains in 2020, Yu sent me photos of Yongxing River Park. The trees and grasses had grown up considerably since I'd seen it two years prior, turning it into a lush, green oasis. The channel contained a good amount of water but was nowhere close to overtopping. Turenscape does not yet have data to measure Yongxing's flood capacity, infiltration rate, or water-cleaning services, but Yu called its handling of the storms a "great performance."

The benefits of earlier Turenscape projects have been measured. Yanweizhou Park in Jinhua City, near where Yu grew up, absorbed a hundred-year flood in 2015, protecting the city. Shanghai's 34.5-acre Houtan Park cleans up to 634,000 gallons of polluted river water daily, improving the water's quality from grade V (unsuitable for human contact) to grade II (suitable for landscape irrigation) using only biological processes.

FIGURE 7.1 Two Turenscape photos of Yanweizhou Park in Jinhua City, China: dry and, during the 2015 monsoon, in flood.
Photos © Turenscape

Urbanization is happening so rapidly in China that whole new cit-ies are being built from scratch. Some of them show what is possible when people plan to accommodate space for water before they start building. Turenscape designed part of the ambitious Wulijie Eco-city in Hubei Province, preserving the natural wetlands for catching and cleaning stormwater on site. This approach reduced construction costs by eliminating the need for underground drainage pipes while conserving habitat for wildlife and vegetation. Buildings have roof gardens and living walls, and pedestrian and bike paths thread through the green space, enhancing quality of life for residents.

Inviting Water In

As in Chennai, it can be challenging to make space for water in a densely populated, already built city. But historical maps show that, over time, buildings come and go. In the United States, many are replaced after fifty years. In rapid-growth places like China, that turnover may happen in just fifteen years (although the average is about thirty years). That presents enormous opportunities to redesign buildings and urban systems to scale up nature-based processes as needed and as urban remodeling allows. Disasters can also serve as a catalyst for urban change, such as when governments use emergency funds to buy victims out of a floodplain, remove buildings, and convert the area to an absorbent park.

Reclaiming derelict industrial sites along rivers can also clear a lot of important space for water. Turenscape oversaw such a project in the thousand-year-old city of Kazan, Russia, which surrounds three oxbow lakes on the Volga River. After the Soviet period, pollution had killed nearly all life in the lakes. The city was also prone to flooding because dams on the river elevated reservoirs above the lakes, and seven pumping stations could not keep up when waters rose.

Turenscape's design for Kazan reclaimed land for floodwater over seventy-four acres, traversing 1.4 miles along the river and its tributaries. The city built linear parks, promenades and bioswales that slow, absorb, and clean urban runoff before releasing it into the lakes. Walking and biking routes give people access to the riparian zone along the river and support human-powered transportation throughout the city. The parks opened in May 2018 and attract thousands of people daily to the formerly deserted waterfront, according to World Architects, an industry group.

In addition to these larger-scale projects to claw back space around natural waterways, other tools of the trade can accommodate water in compact cities. These include bioswales, retention and infiltration ponds, parks, and seepage and injection wells. To work best, these interventions should be linked together so the water can travel some approximation of its natural path and find places along the way to sink into the ground. The idea is to mimic nature as much as possible. Where human space is nonnegotiable, designers sometimes use surrogates, such as permeable pavement and green roofs that can absorb water.

Yu has converted his house, a duplex, into a living laboratory of some of these techniques. Between the two apartments, Yu shows me a living wall he built of porous limestone. Water captured from the roof dribbles down its face, from which maidenhair ferns and philodendrons sprout. The green wall cools the two homes enough that they do not need air-conditioning, although he concedes that it gets a bit warm in summer.

Plants on decks off the bedrooms are watered with roof-caught rain, stored in tanks under the raised plant beds. Yu's deck smells great, emanating whiffs of rosemary, lemongrass, and Chinese chrysanthemum. It even has a tiny creek in which goldfish swim. His sister lives in the other apartment. Her deck has terraced beds replete with lettuce and chard. "We collect fifty-two cubic meters of stormwater [annually], and I grow thirty-two kilograms of vegetables," Yu says proudly. His efforts reduce runoff from his building's roof and decrease his personal water usage from city sources.

Custom Solutions Needed

Like all Slow Water projects, from Yu's deck gardens to Turenscape's biggest transformations, each design must factor in local climate, soil, and hydrogeology. Consider two Chinese cities with diametrically opposed water needs. Kunshan, in Jiangsu Province near Shanghai, is built on polders, land reclaimed from water with levees. The water table is so high that infiltration isn't possible, but filtration—cleaning the water—is necessary. Hotan, a desert city in far western Xinjiang Province, gets barely a half inch of rainfall a year, on average, so flooding isn't an issue, but it needs to protect its groundwater supply.

If China ignores this specificity, its broad ambition for Sponge Cities may falter, says Zevenbergen, the Dutch professor. The rush to develop cities in the past twenty years did not allow builders time to understand imperfections in design and make adjustments. That's what led to cities across China experiencing the same problem at the same time: widespread urban flooding. Rushed implementation of Sponge Cities could also lead to missteps. President Xi's program has strict deadlines, which may not allow time to monitor performance, adjust if necessary and transfer knowledge. It "takes time to learn and to reflect," Zevenbergen warns.

A paper written by Chinese government research institutes in 2017 expressed similar concern about a cookie-cutter approach. It noted that training, long-term data, and new design codes are needed to avoid creating a pattern of problems. To provide guidance, the government has formed the Sponge City Technical Scientific Committee, including civil engineers, economists, and landscape architects, including Yu.

Although Zevenbergen expects the Chinese will make a lot of mistakes along the way, he thinks that "in the end, they will become the leaders in Sponge Cities. The same happened in the realm of renewable energy." That's because the country has a culture of getting things done. "Every year I'm in China, I make designs with students, and the next year the projects have been implemented. And that is really astonishing."

The flip side of getting things done quickly, however, is a culture with less devotion to maintenance and aftercare. And green infrastructure requires a different kind of maintenance than gray, such as pruning or replacing plants, just as the engineer Narasimhan pointed out in Chennai. A Chinese–European, peer-learning exchange program with which Zevenbergen is involved is helping to accelerate the pace of learning.

China needs to learn fast. Gray infrastructure struggled during recent heavy summer monsoons, which pushed several giant dams to the brink of failure and killed at least 219 people. Meanwhile, the Yangtze area alone is choked with so many dams that 333 rivers have dried up to varying degrees.

Zevenbergen, in obvious frustration, calls the massive dams examples of "stupid infrastructure." Giant gray-infrastructure projects like these dams are unlikely to have long lives in the age of climate change because they can take a decade to build and are engineered to accommodate certain maximum flows. To do that effectively "means that we need to know how much climate change we can expect," Zevenbergen tells me. "The problem is, we do not know."

Yet China is still dam oriented. Despite the national promotion of Sponge Cities, Yu says that China's schools continue to train engineers using twentieth-century principles. And in the offices where decisions are made, Yu still confronts a penchant for stronger dams, bigger pipes, and larger stormwater storage tanks—a refrain I hear repeatedly from water detectives around the world. Yu says poignantly, "We are fighting so hard to try to get people to think in an ecological way."

Working Landscapes

Shifting the dominant culture to adopt a new ethos of water and land management is a tall order. During the Industrial Age, people became accustomed to water looking and behaving in a certain way, tidy within its concrete constraints. But water is not a waste product to be quickly whisked away; nor is it a commodity, lying inertly in a reservoir until needed on fields or in apartments. Water has agency.

The cycles of nature can look sloppy to the industrial eye. Yu is trying to expand our aesthetic so people can once again appreciate an area's changes from season to season—form following the function of natural systems. His designs often have a pop of color, such as an undulating red walkway over wetlands that are sometimes flooded or muddy, or benches set among wild plants. These elements invite city dwellers into the dynamic landscape and provide a unifying polish, like tying a bow on nature's constant evolutions.

He speaks widely on his concept of "big-feet aesthetics," that counter-analogy to foot-binding, a sign of aristocracy and beauty because the tiny feet left women unable to work. Urban landscapes filled with nonnative plants that require regular applications of water, fertilizer, and pesticides are like those tiny feet: not useful, the antithesis of resilience. Land and water with big-feet aesthetics, on the other hand, are strong, ecologically healthy, low maintenance, and productive. Some of Yu's designs even produce food for city dwellers, such as Quzhou Luming Park in Quzhou, Zhejiang.

At President Xi's directive, Sponge City landscapes are also working to fix another water problem caused by industrialization: pollution. Nutrients such as nitrogen and phosphorus, along with heavy metals, pesticides, and microplastics, taint surface waters in China, according to Randy Dahlgren, a scientist at the University of California, Davis, who specializes in soil and water chemistry and has worked in Zhejiang Province. "If they can get this water to infiltrate into the ground, a huge number of these potential contaminants will be retained within the wetlands systems, buffers, detention basins and bioswales," he tells me.

But it isn't quite so easy. Human-constructed wetlands in polluted urban areas can't just be built and left to their own devices. Heavy metals and excess nutrients can be absorbed by plants, but they return to the soil when those plants die. "You really need to be harvesting those plants," says Dahlgren. They can be made into biomass fuels and incinerated, although some pollutants such as metals accumulate in the ash, which then requires disposal. Planners should also be cautious about moving surface-water pollution into groundwater, where impurities could persist for anywhere from tens of years to several centuries. This complexity raises the bar for doing Sponge Cities right.

Nevertheless, Sponge City techniques are already reducing pollution in places such as Philadelphia, where they are called low-impact design. Like many US cities in the Northeast and Midwest, Philadelphia routes stormwater through sewage treatment plants. Initially that seemed like a good idea—giving stormwater an extra cleaning before returning it to a river—until urban sprawl led the systems to overflow during big storms, pushing untreated sewage into rivers. With its Green City, Clean Waters initiative, Philadelphia is reclaiming land along the banks of local creeks and rivers as parks to absorb excessive rainfall and flood when necessary. One project brought to the surface a stretch of creek that had been buried in pipe and reclaimed natural space alongside it. Bioswales along the city's zoo parking lot collect runoff from the nearby avenue. The city also gives incentives to landowners to open natural water pathways with street tree wells, planted "bumpouts" in sidewalks, rain gardens, green roofs, urban farms, and porous pavement. Stormwater that percolates into the ground through these interventions reduces the volume entering the sewage system. As of June 2021, the city had installed more than three thousand green stormwater systems at more than 850 sites, reducing the combined sewer overflow volume by more than two billion gallons annually.

Slow Water projects in cities also improve human health in other ways. That's because they make space for nonhuman life, and humans are biophiliacs, according to the influential biologist E. O. Wilson. That means we are biological beings who love other biological beings. We love life. Many, many studies have shown that people suffer increased rates of depression and anxiety when they are separated from nature, and experience a calming effect when they are reconnected. Spending time in natural areas—even urban parks—improves focus for kids with attention deficit hyperactivity disorder, enhances mood for people with depression, and helps patients recover more quickly from surgery. There's some evidence that bacteria in soil is an antidepressant—at least in mice, for whom they increased serotonin levels. Forest bathing, a popular pastime in Japan, has psychological and physiological effects. Compounds emitted by trees boost the immune system and provide relaxation, according to botanist and medical biochemist Diana Beresford-Kroeger, author of *To Speak for the Trees* (2019) and *The Global Forest* (2010). It makes sense. We are animals. We evolved in nature, our original habitat. That's why our degradation of the natural world harms us both physically and emotionally.

Biodiverse Lands

The most effective way to restore degraded landscapes—for human health, to support struggling species, for Slow Water benefits—is usually with native plants. That's Turenscape's approach. It uses mostly native plants because they are adapted to the local environment and need no supplemental water, fertilizer, or pesticides.

Although using native plants in gardens and urban landscaping is growing in popularity, it's still far from mainstream. As people have spread across the globe, they've purposefully and accidentally taken other species with them. They bring favorite foods, or plants that remind them of where they grew up, or a species meant to "solve a problem," such as constricting beach dunes in California with South African ice plant. As a result, we have remarkably similar plant species in cities the world over. But just as water engineering has obscured from us what water wants, these plant species from elsewhere separate us from our places' natural heritage and identity.

Imported plants often have an outsized impact on water resources too, as did the problematic *Eucalyptus*, native to Australia, when planted in Peru. Lawns are among the biggest water villains. The largest irrigated crop in the United States, lawns suck up nearly nine billion gallons of water per day. We also douse them with chemicals, killing soil microbes and sending polluted runoff into waterways.

It's worth thinking about what we're losing to lawns. My friend Maleea Acker is a geography professor and a native plant activist in Victoria, British Columbia. She wrote a book about First Nations' stewardship of the landscape prior to colonization called *Gardens Aflame: Garry Oak Meadows of B.C.'s South Coast* (2012) and has converted her suburban yard to a native plant landscape. She has written elsewhere about her motivation: "When we try to make something into what it isn't (a lawn is a nostalgic memorial to England's sprawling estates), we disconnect from what is actually here: moss, licorice fern, fairy cup lichen. . . . It's time to jettison these damaging preconceptions. Time to live in place, where we are, not some tidied-up version of suburban glory. Let's bring the beauty of our parks home, so that other species can also live outside those refugia."

I have seen biodiversity increase in heavily paved San Francisco, where the houses are flush up against each other. The yard behind my flat is a rectangle about 25 feet by 50 feet. I wanted plants that wouldn't need supplemental water, so I started researching natives because I figured that, if they evolved here, they could take what they need from the rain and fog that comes their way. I learned from the local chapter of the California Native Plant Society that San Francisco has a special biome that is nearly gone because the city is so densely developed. But San Bruno Mountain south of the city has been protected by conservationists, and plant experts from the society harvest seeds from plants there to sprout and distribute. Due to all the concrete in the city, people's yards have become refuges for animals. Since ripping out English ivy and crabgrass and filling my yard with ceanothus and manzanita, I've seen red-tailed hawks, egrets, a great blue heron, raccoons, carpenter bees, Steller's jays, California slender salamander, and much more.

While renaturing parts of the city to make space for Slow Water is unlikely to recreate a fully intact ecosystem with all its original critters, as we saw with Seattle's Thornton Creek in chapter 3, it will create more habitat than we have now. It really works: I have collected a fat file of stories about people noticing species returning after they restore a bit of landscape.

Human health, too, benefits from increasing biodiversity in our cities. That's because chronic diseases such as asthma, inflammatory bowel disease, and allergies have been linked to less-diverse human microbiomes—the microbes that live in and on us. Replanting urban environments with native flora helps rewild and increase diversity in the environmental and human microbiota, researchers say.

As we learn more about these benefits, our tastes for what's beautiful can change. The industrial farm and city landscapes we've created don't have to be our legacy. We can set back levees and make room for floodplains and wetlands. We can create urban green spaces full of native species that also serve what water wants. We just need a postindustrial vision.

A couple of years ago, a mutual friend encouraged me to meet

Toronto-based Anishinaabe filmmaker Lisa Jackson. Her work is wide ranging, but Indigenous stories are a regular theme. Jackson is witty, well read, creative, and open; I liked her instantly. She was in Victoria, BC, my part-time home, to show her latest work, *Biidaaban: First Light* (2018), created in virtual reality. I'd somehow evaded trying VR until that day. I put on the headset, which covered my eyes. In an instant, I am standing in a future Toronto. I can take small steps forward or back within the scene and move my head to look around.

Biidaaban: First Light is a bit like *Blade Runner,* in that it portrays an urban futurescape that is at once familiar and foreign. But *Biidaaban's* Toronto is the opposite of a hellscape populated by desperate people. Instead, I'm standing on the flat roof of a building, looking down on a peaceful city square in flood. Native plants are edging their way back in amid the concrete, and behind me, Indigenous people have built a traditional structure, rounded, organic, womblike. Wendat (called Huron by settlers), Kanyen'keha (Mohawk), and Anishinaabe (Ojibway) words float by, carrying traditional knowledge of human collaborations with nature, demonstrating the vital familiarity with this place that is embedded in local languages. Jackson's vision of the future is recognizable but more holistic than today's reality. Rather than replacing nature's busy patterns with concrete control, or embracing technology as savior, it centers on community—with people and other beings. It invokes the Indigenous roots in this land, Indigenous values of gratitude and care for the natural world that sustains us.

Suddenly, the cityscape drops away from under my feet, and I gasp because it feels like I'm falling. But no—I am supported, standing in space, stars arrayed around me. I am part of the universe. We all belong here: people, other animals, plants, water. That knowledge, that inclusion, that functionality and resilience surrounds us still, if we are willing to open our minds to consider things ancient, new again.

· 8 ·

For Future Humans

PROTECTING WATER
TOWERS IN KENYA

If you destroy the forest then the river will stop flowing,
the rains will become irregular, the crops will fail and
you will die of hunger and starvation.

WANGARI MAATHAI,
winner of the Nobel Peace Prize

In Kenya's Aberdare National Park, Peter and I head to our quarters at Treetops Lodge: the Princess Elizabeth Suite. The room honors England's longest-serving monarch and current queen, who was staying at Treetops in 1952 when her father died, leaving her the throne. A couple of years after her stay, Mau Mau freedom fighters torched the lodge's original treehouse design, featured on the popular Netflix drama *The Crown*. This is a rebuilt, expanded version. We are given the marquee quarters because it's February 2020. Covid-19 hasn't yet arrived in Kenya, but it's already locked down Asian and European tourists, and we are the only guests. It feels a little freaky, like a *Scooby Doo* episode in which the gang goes to a seemingly abandoned hotel, but we quickly succumb to the comfort and hospitality.

Fortunately, the animals are not staying away. The main draw at Treetops is a pond directly at its feet, where wildlife come to drink and bathe. Groups of elephants gambol by, Cape buffalo wade in from the other side of the pond, Thompson's gazelles make their skittish appearance. The hotel has a buzzer system to alert guests in the night to certain animals: one buzz for hyenas, two for leopard, three for black rhino, four for elephants. Our buzzer sounds three times our first evening. Rhino! We rush to the bay window and peer into the deepening night. Squinting through binoculars, we can just make out the animal's beefy shoulders on the far side of the pond and—as he turns—his horns. Because rhinos are so endangered, it almost seems like a brush with a unicorn.

Aberdare is a refuge in modern Kenya for these animals and many more. A 249-mile-long electric fence around prime forests and water catchments are meant to keep the wild animals in and people out— especially those who might cut trees, plant crops, or hunt illegally. Around Treetops the landscape is grassland flocked with deciduous trees. Hyenas had killed a buffalo along the dirt road on the night we arrived and, over the next thirty-six hours, we watch the carcass disappear in an ongoing hyena party. The park, with its relatively intact ecosystems, is not just a haven for more than fifty species of mammals, 270 types of birds, and 770 kinds of plants. That richness makes it a draw for one of Kenya's leading industries—tourism.

Even more important to many Kenyans, Aberdare National Park is the primary source of water for Nairobi, Kenya's capital. The Tana River originates here, and two of its tributaries, the Thika and the Chania, supply the city. The Tana River basin also provides about half the country's hydropower. The Aberdares, with elevations soaring to thirteen thousand feet, is a "water tower"—meaning a highland source of water. As in Peru, most Kenyans rely on mountains for their water. The Aberdares is one of the five biggest water towers in Kenya, a group that also includes the Mau Forest, Cherangani Hills, Mount Elgon, and Mount Kenya, whose elevations range from about ten thousand to seventeen thousand feet. Mountains provide water because when warm air hits the side of a peak and is forced to rise, it cools and can no longer hold as much water vapor. That excess turns to liquid droplets, forming clouds. When those clouds release their rain, the plants and trees that cloak mountains help slow precipitation on slopes, and the space in soil carved by roots creates pathways for water to move underground. Eventually this water supplies rivers like the Tana that flow to the cities and towns below. Plants and trees also pull water up from the ground, take what they need, and release water as vapor into the atmosphere through tiny holes in their leaves called stomata. This process, transpiration, is similar to the way animals release water vapor with every exhalation. A single tree can transpire hundreds of liters of water a day, so forests play a critical role in supplying rain both locally and farther away.

Scientists are still exploring how healthy forests support a healthy water supply. But in Kenya, it's conventional wisdom that deforestation leads to declines in water flow because the phenomenon has been observed directly by local people. Nobel Peace Prize winner Wangari Maathai was a social and environmental activist, the first woman professor at the University of Nairobi, a member of Parliament, and a minister of Environment and Natural Resources. In the late 1970s, women told her their forests were disappearing to firewood as well as government-sanctioned logging and tree farms. As a result, their streams were drying up, making their food supply uncertain. In response, Maathai founded the Green Belt Movement and engaged women, mostly, to reverse these negative trends by replanting more than fifty-one million indigenous trees, primarily around the five main water towers. (Kenyans use the term *indigenous* rather than *native* for plants.) She also fought against agricultural encroachment and government land grabs of forests.

Today, finding ways to keep that water moving through mountains rather than running straight off, via forest conservation and sustainable agriculture, is critical to water security here. Kenya is almost as large as Texas and ranges from a tropical climate along its Indian Ocean coastline to an arid northeastern interior to a fertile plain to the west. It's location at the equator makes parts of it lush and verdant, especially after rains; yet much of it lies at high elevations, where the climate is comfortable. Its central Highlands region is bifurcated by the Great Rift Valley, a product of plate tectonics that is slowly splitting Africa apart. The country is prone to regular floods and droughts, which threaten food, water, and energy security. (Kenya gets 34 percent of its electricity from hydropower dams. Across the continent, one-third of nations get the majority of their power from hydroelectric plants, and droughts are making that supply unreliable.) Kenya's major population centers are the capital Nairobi, which sits on a temperate high plateau at about six thousand feet and is home to six million people, and Mombasa, home to 1.2 million on the hot, humid coast.

Protecting the Aberdares, Mombasa's Shimba Hills, Eldoret's Cherangani Hills, and the other water towers has become more difficult in recent years. Kenya has seen additional pressure on water supplies from population growth, increasing consumption by the rich, and international resource capture. The latter can be legitimate business deals between governments and international investors. But there's also a global phenomenon of land grabbing, in which rich countries and corporations are increasingly buying or leasing large swaths of land in poorer countries for food and water. In the case where a rich, dry country is using the land to grow food or products it couldn't grow at home, it's essentially a water grab. The difference between a deal and a grab is that the latter runs roughshod over environmental protections and human rights, such as forcing people off their traditional lands to hand it over to corporations. An independent global land-monitoring database that tracks land grabs has found them taking about 1,350 square miles in Kenya.

But population growth and the struggle to provide adequate food, water, education, and health care to people are the factors that preoccupy all the government officials I met in Kenya. The country's population has grown from sixteen million in 1980 to fifty-four million today. In the last few decades, people have increasingly moved into highland forests, cut trees for charcoal, and planted crops. These activities have constricted space for wildlife and made farming more difficult because, without trees to hold the soil, it runs off the mountains when rains come, washing away nutrients. And because the land is not holding the water and no longer has trees that help to form rain, ultimately streams dry up for want of water through the underground. Downstream, water is laden with sediment and needs more-expensive cleaning, wet seasons have more prevalent flooding, and dry seasons see more water scarcity and hydropower cuts. At the same time, climate change is bringing longer droughts and bigger storms and floods both in Kenya and throughout Africa; the continent is projected to warm up to 1.5 times more than the global average, according to the United Nations.

The population growth in Kenya is a piece of what's happening broadly across Africa, where a population boom is underway this century. The size of the continent is difficult to grasp. Its fifty-four countries span an area almost twice the size of Russia, or more than three times the size of the United States. Today Africa's population is 1.3 billion people. Compared with, say, India, where more than 1.4 billion occupy just one-tenth the area, that means there's still a lot of room for other species. That Africa still has so much of its natural landscapes, including more free-flowing freshwater systems remaining than any other continent, is why its large mammals—elephants, giraffes, cheetahs, lions, gorillas, chimpanzees, rhinos—still survive here, for now.

But while human population growth is slowing in many places around the world, numbers across Africa are projected to double in the next thirty years, to 2.5 billion, and to surpass 4 billion by 2100. Population growth remains a sensitive issue in many countries, with fears of bias against religion, culture, poverty, or race, but in fact it's largely a reflection of women's disempowerment. A study in Kenya found that the number of girls reaching secondary education increased from 12 percent to 59 percent between the mid-1970s and the mid-2010s, while contraceptive prevalence grew from 5 to 51 percent and births per woman dropped from 7.6 to 4. Yet 40 percent of women in sub-Saharan Africa still lack access to reproductive health care, including contraception, and are unable to make choices about the size and spacing of their families.

This increase in human population, as well as growing industrial development, golf courses, and the like, is also increasing demand for water. Yet roughly 25 percent of people across the continent already suffer from water stress. As African leaders grapple with how to provide for more than three times the number of people who live here now, they have an opportunity to sidestep the water management choices that are causing problems elsewhere and find a better way. They aim to provide for growing populations while protecting the continent's natural wealth and wildlife. In Kenya, that transition is underway. Government and NGO leaders are recognizing the connection between ecosystems and water in policies and institutions. Some decision makers are beginning to eschew concrete water infrastructure and instead protect forests and high-altitude grasslands, the water generators. They are conserving lower altitude wetlands, which are water storage facilities. And they are supporting farming practices that capture rainfall, conserve water, and keep soil on the land rather than letting it run off into rivers—strategies that also increase crop yields. Here, water detectives don't have to consult history to figure out what water wants because the change from natural to human-altered water systems is happening now, and within living memory.

Tower Power

Our second day in Aberdare National Park, Peter and I set out to visit the grassy moorlands at the very top of the water tower. At about eight thousand feet elevation, we stop at a cabin to pick up a young ranger. He is required as a safety chaperone if we wish to step outside of the car and hike due to the wild animals, some of whom could, conceivably, eat us. Like most rangers in Kenya, he is dressed in camouflage and carries an automatic rifle slung over his shoulder.

"My name is Joshua," he says.

"My brother's name is Joshua!" I reply.

"I am like your brother," he smiles, and I can feel the truth of it.

We drive down dirt tracks, nearly overgrown with thigh-high grasses. As we gain in elevation, we move through different ecosystems. First there are mixed montane forests, where we see olive baboons crossing the road, the leader stopping for a moment to stare us down. Then the path winds through endangered rosewood forests, the trees' gnarled branches held above us in an open canopy like ballerinas' raised arms. The trees' beautiful red wood, cloaked in thick moss, is popular for furniture in China, and Asian rosewoods have been nearly poached out. East African rosewoods are also under pressure, which is why these charismatic trees are so rare.

FIGURE 8.1 Atop Aberdare National Park lie the moorlands, headwaters of the Tana River that supplies much of Nairobi's water. Peter Fairley and Joshua Kemboi hike. Photo © Erica Gies

Regretfully leaving them behind, we move into towering native mountain bamboo that sways in the wind and leans over the road. Here, grayish brown-and-white Sykes' monkeys tumble over each other and cling to stalks, checking us out, while eastern Colobus monkeys scamper past, dapper with their long fringes of black-and-white fur and odd faces.

Eventually we emerge above it all, in the high-elevation moorlands that start at around eleven thousand feet. The landscape is enchanting, studded with waist-high tuffet grasses that ripple in the wind, yucca-like plants called giant groundsel, and Seussian giant lobelias with green flowers that look like massive pine cones. The mountain peaks are little knobs of rock that poke up out of the rolling ground. As we start walking, I discover that parts of the trail are muddy, and in lower spots between the undulating hills, water seeps. Bogs and mountain streams that form here and elsewhere in the Aberdare mountains, park and forest, feed the rivers that supply four of Kenya's six largest drainage basins.

The vast scale of the landscape and its smorgasbord of unusual, tiny alpine plants are exhilarating. Joshua, whose last name is Kemboi, knows many of their names, which I appreciate. Each time I stand up after squatting down to examine one closely or take a macro photo, my head whirls due to the low oxygen. Although leopards have been spotted here on occasion, the most dangerous critter we see is a montane viper sunning herself on the trail. The snake is miniature, like the plants, but her bite is deadly.

"How fast does the poison act?" I ask.

"Pretty fast," Kemboi replies.

Peter gestures to the ranger's cellphone. "Do you get reception up here?" he asks.

"No," Kemboi says, and we silently contemplate our isolation and the multihour drive to a hospital for antivenom. "Better to avoid getting bitten," he adds matter-of-factly, leading us well off the trail and around the small snake, who seems perfectly content to continue her sunbath.

*

The national government recognizes the importance of protecting these ecosystems atop Kenya to maintain water security. In 2012 it created the Kenya Water Towers Agency to help do that. It's one of several forward-thinking policies and government departments created to manage water in Kenya. The 2016 national Water Act introduced and strengthened other innovative approaches. One, the Water Resources Authority, allocates water proportionally based on how much is available in a given year. That may sound like common sense, but it's not standard practice everywhere. For example, in California and much of the western United States, water officials allocate set quantities of water to the users who made the earliest claims in the state's history. That means in a dry year, those preferred users still get all their water while newer users might go entirely without; not everyone "shares the pain." Another savvy Kenyan policy strives to involve local people in the responsibility of caring for their local water resources. Community Water Resources Users Associations get together to make decisions about their local water management. However, finding funding to support these organizations is an ongoing challenge.

These unusual agencies and policies are the fruit of a government realignment in Kenya. After independence from Britain in the 1960s, Kenya was ruled for thirty years by two dictators, Jomo Kenyatta and Daniel arap Moi, who continued a form of colonialism benefiting themselves and their cronies. But in 2004, Kenya asserted a second stage of independence, with new leaders and the beginnings of a constitution written by Kenyans, rather than one propagated from British Parliament. Kenyans were also motivated to make these changes in water management partly in response to droughts that cut electricity and exacerbated food insecurity.

"Before the reforms, water access was becoming very difficult," says Boniface Mwaniki, technical coordination manager at the Water Resources Authority, who I meet in his office in Nairobi. Kenya ping-pongs between recurrent droughts and severe floods, which interferes with water supply and hydroelectricity, leading to rationing. According to Mwaniki, these cycles are accelerating: "Flood and drought used to be every ten years, then every six years, then three, now almost every year."

On paper and by design, these water policies are among the most progressive in the world. Of course, in Kenya, as elsewhere, the stated goals of government policies and agencies can be slowed by bureaucracy, corruption, or lack of funds. Execution, which happens at the local level, is often difficult. Still, enshrining the holistic stewardship of natural resources in government policies is a notable divergence from the classic western development model that raids natural resources for capital.

Other countries could also benefit from expanding natural ecosystem protections for water security. For example, New York City famously bought 1,600 square miles of land in the Catskill Mountains for $1.5 billion in 1997 to protect and clean its water, rather than invest in an expensive new water treatment plant that it estimated would cost more than $6 billion plus $250 million a year for maintenance. It was cheaper to protect the forest. Often, conserving watersheds can be less expensive than destructive alternatives to supply water, such as desalination plants or new reservoirs formed behind dams.

Africa's First Water Fund

Aberdare National Park is the crown of the Highlands. Just downhill, the protected area is ringed by Aberdare Forest Reserve, and below that, small family farms begin. In recent decades their number has burgeoned to three hundred thousand, and farmers have cut forests that used to stabilize the soil and water.

Downstream, where city dwellers depend on rivers flowing from the towers, water insecurity is growing, especially for people living in large slums, such as Nairobi's Kibera settlement. Only 53 percent of people in the capital have direct access to water, according to the World Bank. And the people who can least afford it pay the most: those who buy water with e-money from stations in the slum or from trucks, which is often sourced by pumping groundwater, pay up to fifty times more than city water rates.

To help support government agencies in their progressive goals for water management, the Nature Conservancy (TNC) introduced the Upper Tana–Nairobi Water Fund in 2015. The fund links farming and forestry practices in the Upper Tana River watershed to water flow and quality in Nairobi. It is similar to the water funds TNC helped pioneer in South America, including those in Colombia and Ecuador that inspired Peru to create its Mechanisms of Reward for Ecosystem Services program. In Kenya, TNC has raised more than $20 million, including about $1.6 million from private companies that use water, such as Coca-Cola and a brewery. Government agencies are also supplying funding or institutional support, including the national Water Resources Management Authority, the Nairobi City Water and Sewerage Company, and the electricity utility KenGen. Interest on the water fund's principle is a stable source of funding for projects that help maintain the natural function of watersheds.

Water-fund projects in the Upper Tana focus on two main areas. On national forest land, they are replanting indigenous trees. And just below that, where the small family farms begin, projects encourage farmers to implement water- and soil-friendly farming practices. These strategies help farmers to slow water on their land, where it lingers longer in the soil, helping them to produce more food, raising their incomes. The interventions increase groundwater recharge to keep rivers flowing in dry seasons and reduce peak flows that might flood areas downstream. They also lessen the sediment that would otherwise need to be removed regularly from gray infrastructure, raising treatment costs, because it decreases hydropower production and water quality.

A local leader who helped found the water fund is Fred Kihara, Africa Water Funds director for TNC. I meet him in Nairobi at the organization's office, a polished compound in a northwestern residential neighborhood. His family's history is part of the development of the Upper Tana watershed. A middle-aged man with a nearly shaved head, wearing a button-down shirt and a beaded bracelet in the pattern of the Kenyan flag, Kihara grew up just on the edge of Aberdare Forest Reserve.

"The history of farming is not old here," he explains. Instead, Kenyans in the area mostly kept sheep and goats. To provide for their animals, "people would keep good pasture fields. And they would also take good care of swampy wetland areas because those remained even when it was dry. They were places to water your animals and draw water for households."

After Kenya's independence from Britain, Kihara's parents got a small piece of land and began tea farming to sell the cash crop to their former colonizers. As he helped his parents with the tea and livestock, he noticed that farmers' success was eroding literally and figuratively over time, particularly as people moved in to cultivate very steep plots.

"By the water running down the fields, it's carrying their topsoil," he tells me. "They're losing their nutrients. Even if they put on fertilizer, it runs off."

They needed to slow water down.

Kihara went to Nairobi to study soil and water engineering and later became a soil conservation officer with the Ministry of Agriculture in an area near Mount Kenya. He learned how to conserve land via techniques such as terracing and agroforestry. These practices, which retain soil in place and therefore water and nutrients, help farmers reap better yields with less water and fewer fertilizer applications. "The advantage is, even when rains are not good, you still get a good crop," he says.

In the Upper Tana watershed, soil poverty has led to real hardships. For that reason, the most important goal of these projects is to benefit farmers. The Upper Tana–Nairobi Water Fund estimates that it has given twenty-nine thousand households some benefit, such as food security. "We've just planted one million avocado trees," says Kihara. "In three years, the trees will be producing avocados for a better diet and as a product to go to the international market." Other important goals are to increase soil and water conservation on farms compared to the 2017 baseline.

As Kihara describes it, avocado trees also meet these goals. "They can exist on rainfall alone, and they help to hold the soil." Supporting farmers brings important ancillary benefits as well. Healthier soils store carbon, rather than release it. Widespread adoption of sustainable agricultural practices also helps protect wildlife because people's increased food security means they are less motivated to encroach into wild areas or to poach wildlife to eat.

So far, improved soil management practices have been adopted across fifty thousand acres of the Upper Tana River watershed, and there has been a 40 percent reduction in sediment load compared with 2013 in key rivers supplying Nairobi. TNC scientists also measured nearly five million more gallons of water per day, on average, flowing into the main Nairobi reservoir in 2020 compared with 2016 levels.

The Upper Tana–Nairobi Water Fund was the first of its kind in Africa, but its model is spreading. There's already another operating: the Greater Cape Town Water Fund in South Africa. Plans for additional funds are also moving forward in Kenya in Mombasa and in Eldoret–Iten in the Rift Valley. Other places interested in water funds include Malawi; Uganda; Tanzania; Morocco; Freetown, Sierra

Leone; Addis Ababa, Ethiopia; Beira, Mozambique; Kigali, Rwanda; Dakar, Senegal; and Durban and Port Elizabeth in South Africa.

Farming Smarter in the Highlands

About two-hours' drive northeast of Nairobi on a bright February day, I travel with Caroline "Carol" Nguru, a Murang'a County agricultural extension officer, to visit farmers who are improving their practices. Driving around the watersheds of the Thika and Chania Rivers, we pass through villages that each have just a few streets, with buildings made of concrete, stone, mud and grass, or corrugated metal. Hand-painted signs advertise churches, schools, and shops selling such things as meat and milk. This area, called the Highlands, ranges from 4,500 to 8,000 feet of elevation. The scenery of rolling hills and steep mountains unfolding before us is idyllic, with its intense blue sky, red soil, and lush green vegetation, dotted with people in colorful clothes.

Driving in Kenya is very slow—cars rarely surpass thirty miles per hour—due to dirt roads, potholes, and many speed bumps, some of which are quite tall or totally unmarked. Slowing cars is the point because roads are also busy with people walking, or hanging out on the grassy margins, talking. For greater distances, many people ride *matatus*, large minivans filled with fourteen people and stuff strapped on top. Others have their own motorcycles, or motos, and carry all kinds of things tied behind, from water jugs to a full-sized dead pig.

Tea, the primary cash crop in this watershed, blankets the steep hills in tidy, waist-high squares of green. Between them are tiny paths so workers can pick leaves daily, thrusting them into baskets a couple of feet tall and as big around as curved arms. When full, people carry the baskets on their backs, strapped across their foreheads or chests, to roadside pickup points. At one spot on the road, the number of people carrying tea baskets ticks up, and we come upon a big truck with an open framework in the back, studded with hooks. People line up, tipping the contents of their baskets into canvas bags that are then hung from the hooks via straps.

Nguru works with the Nature Conservancy as a water-fund liaison. She travels around this watershed, meeting with farmers and encouraging them to implement better practices that reduce soil erosion and safeguard water quality and quantity. Today she is taking me on her rounds so I can meet some of them and see their hand-hewn water innovations. In her thirties, with thick braids and intelligent eyes, Nguru exudes an effortless cool. She is dressed in wellies, a Zoo York purple hoodie, and a woven visor. Warm and friendly, she is equally at home educating a foreigner about the arcane policy and science behind this work as she is tromping through thick mud, talking to farmers. With them she is both prescriptive ("This trench you've dug is too high on the slope side") and humble (showing her interest in a farmer's innovation, asking, "What are you doing here?"). She is also liberal with praise ("I'm *so* impressed with this farm"). The farmers seem genuinely happy to see her, showing off what they've done, asking advice, requesting more avocado tree seedlings.

Like the people who live in this area, Nguru is Kikuyu and speaks that Indigenous language. While all Kenyans speak Kiswahili and many speak English, they also speak their own languages; there are at least forty-two across the country. Politics and many other facets of life here follow ethnic lines. That carries over into Nguru's work too. For farmers to listen to her suggestions, she must speak their language, she explains in her low, musical voice.

Success lies in meeting people where they are—with language, culture, and addressing their specific needs. The Nature Conservancy's strategy also includes recruiting people from within the community to tell local people about innovations, like Daniel Muchiri, who rides with Nguru and me for a while and extols to his neighbors the virtues of small storage ponds called water pans, one of the water fund's sponsored interventions.

Nguru puts it this way: in everything we do, we need to think of farmers' livelihoods. They are not paid directly for their practices that improve the watershed. But they are earning more money because their land retains water better and is therefore more productive, or because new profitable crops have been introduced, like the avocados. "If you don't make it beneficial to the local people, if you don't bring them on board with the solution, they won't be invested in it and won't continue it," she concludes.

*

At the first farm high above Wanyaga town, in hills enshrouded by cool fog, we visit Esolom Wandaka Kiguru. Carrying a machete, he is wearing a sweater against the chill and rubber boots. A charismatic older guy, he has a sense of humor that zips out like a whip.

As we walk through the upper portion of Kiguru's land, Nguru points out places where he has planted napier grass and brachiaria, indigenous grasses that have deep roots to hold the soil and provide perennial fodder for cows and goats. Most farms have a cow or two and a few goats or sheep. Animals don't wander to graze in the Highlands because each family's farm is cheek by jowl with another, perhaps one to four acres each. While people sometimes tether sheep and goats by the side of the road for grazing, they generally bring grass to the cows, loading it into trucks or strapping unwieldy bundles to the backs of motos.

Kiguru invites us into his tea fields, and we walk down, down, down tiny paths so steep I have to lean back as I walk. Many farms are on similar grades, which is why it's so easy for soil to wash down into rivers. Across the narrow valley, on the other side of the river separating the properties, another farmer's squares of tea blanket the hills. "Tea is my friend," Nguru tells me, because the plant remains in the soil year after year, holding the soil; people just pick the leaves. That's looking at the bright side of a big shift to intensive cultivation; in Wangari Maathai's autobiography, *Unbowed*, she writes that this area was native forest during her childhood in the 1940s and '50s.

Where the land finally flattens out at the bottom, Kiguru has planted squash and maize, but right along the river he maintains buffer strips of indigenous vegetation, including spiny tree ferns and draping vines. Kenya has a law requiring farmers to retain a twenty-foot strip of natural vegetation along the river to protect water quality, but it's not well enforced. Nguru is obsessed with water quality. She stops the car at every river and stream we pass as we make our rounds to point out its degree of clarity based on the number of upstream projects. In the best cases, the water is nearly crystal clear, like it is here at the foot of Kiguru's farm. At the worst, it boils and dances with red mud.

After hiking back up the steep hill, we leave Kiguru and drive a short distance to another farm. Clouds of African caper white butterflies flutter by, like magical summer snow. White with blackish-gray spots, they travel en masse on regular migrations. Our driver, Dickson Ngure, has a nervous laugh punctuated by favorite sayings: "OKOKOKOK!" or "Wowwowwowwow!" He says his grandmother told him that when the butterflies are moving from west to east, it means the rains are coming. Nguru, on the other hand, recalls that west to east means drought. Regardless of which version of folk wisdom is correct, it's a sign of the importance of the water supply here and the culture of looking to nature to provide.

*

At the next farm, Ruth Kamau, a middle-aged woman dressed in a skirt, shirt, rubber boots, and head wrap called *kilemba*, is also growing tea but is expanding into avocados. Her farm feels like a waking dream, with its trilling birds and sloping green hills, topped by a cerulean sky dotted with big fluffy clouds. She and her husband have dug three depressions for which the water fund has supplied a thick, flexible plastic liner—these are the water pans. Rainwater runs off the corrugated tin roof of her house, through pipes, and into these ponds that average around 8,700 gallons each. We walk down through her avocados and Kamau turns a tap with a satisfied smile as water gushes out horizontally over her tea fields.

The pans are very popular with farmers, according to Craig Leisher, the Nature Conservancy's monitoring and evaluation director for Africa, whom I speak with later via phone at his home in New Jersey. More than fourteen thousand are currently operational. Because there are two rainy seasons in Kenya, the short rains in October and November and the long rains from March to May—with dry periods in between—farmers get two chances a year to collect water. These water pans allow a farmer to grow an extra crop, such as French green beans, during the dry season. "It's a huge bonus on your income," says Leisher. "The farmers love water pans. The best evidence is when people start spontaneously building their own."

Because the ponds are lined, they don't infiltrate rainwater into the soil and groundwater directly. But since farmers use the captured water to irrigate crops, a lot of it does ultimately return to the river. Also, when farmers use water from the pans, they are not taking water from the river; this leaves more flow for aquatic species and downstream users. Plus it saves farmers like Kamau the labor of hauling water up those serious hills and reduces their incentive to plant right alongside the river to avoid lugging water. Leisher is conducting an ongoing study to measure the impacts of water pans on water quantity and quality in local rivers, using electronic sensors downstream of projects, coupled with household surveys. In this way he will be able to compare microwatersheds with projects versus those without.

He goes on to explain that water pans are good for nature too. "The frogs in the Upper Tana love these water pans. There aren't many places in the world where amphibians aren't seriously in trouble. We've created a lot of habitat."

I'm skeptical that these pans, which are a bit like shallow swimming pools, would suit frogs.

"They're very happy with these new accommodations," Leisher assures me. "And you get a lot of birds, both that come through the area and that there live year-round because it's a source of water that wasn't always there."

Tracking Water

Proving such benefits so that decision makers will invest in more projects requires monitoring and studies. But that can be difficult to do, even though the Upper Tana–Nairobi Water Fund earmarks 10 percent of project funding for research. The next day, Nguru and

I set off to another watershed on the other side of the mountain to check out a monitoring station.

Here in the hotter Maragua watershed, coffee and bananas have replaced tea. Just outside the town Irembu, we pull off the road next to a fast-moving stream that winds through tall grass. A river acacia tree captures my attention, decorated with about fifty bird nests that dangle from its branches like Christmas ornaments. The nest makers, lesser-masked weaver birds, are out and about. Sharply dressed in yellow, the males with black masks, they flit about, chittering. On the ground a few nests have blown down. I pick one up. It's an amazing feat of crafting, especially for beings without opposable thumbs. Made of wide grasses, it is indeed woven into an oval sphere with a perfect round hole for the entrance.

Nguru calls me back to the water lesson, pointing out a gauge strapped to a metal pipe set in the stream, which is just a couple of feet across. The device is one of twenty-six data loggers scattered across three watersheds that, since 2015, have recorded water levels every thirty minutes. Someone goes out once a month to collect the data.

As I learn later from Leisher, all the recorders are placed in front of culverts such as pipes where the width of the stream is constant. With that variable stable, you can measure the water level and calculate how much water is flowing by. In addition, the water fund has six more-expensive flow monitors that send data telemetrically every two hours to the Nairobi water supply treatment facility, "so they can see when they've got a plume of sediment or spike of water coming their way," Leisher explains. The devices are too costly to have more, plus they are targets for theft because they contain a cellphone card with unlimited minutes. For that reason, the flow monitors are not necessarily sited optimally to answer study questions; rather, they're set up in fortified areas where someone can keep an eye on them.

Tracking the potential amount of reduced sediment and increased water availability is a long-term process, Leisher says. It takes a minimum of three years to measure rainfall averages because the swings from year to year are "dramatic." He hopes to publish a five-year study in 2022. "I think water quantity is going to be the big thing," he tells me. Early results look like quantity is increasing as a result of the farmers' new practices. In areas where water pans have been adopted intensively, water quantity has increased by more than 20 percent compared to areas without pans.

Forests Jump Back

The butterflies are still snowing as Ngure, the driver, and I say good-bye to Nguru. We leave her watersheds and head to Nyeri town, just outside Aberdare National Park. There we pick up agricultural officer Sabina Kiarie, a middle-aged woman with short hair and a deep-creased smile. Her purview is the Sagana–Gura watershed.

Above the farmlands, the high-elevation forests have been under assault from encroaching farmers and hunters and government-sponsored tree plantations. Since at least the 1980s, groups such as Maathai's Green Belt Movement have been replanting trees and trying to protect those that remain. In recent years, the Green Belt Movement has focused more intently on watersheds than on political boundaries. Its leaders increasingly understand how deforestation—or planting trees—can have long-distance impacts on rivers and underground water flow.

Now the Upper Tana–Nairobi Water Fund is also supporting forest restoration by planting indigenous trees in protected forest areas. "Tree species are chosen because they're adapted to altitude, rainfall and soil," explains Kihara, the Nature Conservancy director in Nairobi.

Sabina Kiarie is taking me to see these returning forests. We stop at a roadside pullout overlooking a valley, with a view of what appears to be a forest primeval, lush and green. Below us is the confluence of the Chania and Zaina Rivers, and a waterfall on the Zaina tumbles down in the distance. This 712-acre area is called the Gakanga restoration site and lies within Aberdare Forest Reserve.

Just a decade ago, all of this was deforested, Kiarie tells me, by people who moved in to farm and cut trees and other vegetation for heating and cooking. Burning plants for these uses is common across much of the continent: biomass supplies almost half the energy used in Africa, leading to some of the world's highest rates of deforestation. Many projects are working to decrease that figure, from connecting people to electricity to introducing the use of solar stoves.

The restoration at Gakanga, begun in 2014, was a collaboration among the Green Belt Movement, Kenya Forest Service, Community Forest Association of Zaina, and Chania Water Resource Users Association. Since the work was completed, seventeen streams that had gone dry have returned, Kiarie says proudly. Near where we stand, purple flowers bloom and butterflies dance. The project is part of broader reforestation work across the Aberdares.

The people who had been farming here were moved out in 2009. Winifred Musila, a director of ecosystems assessment, planning, and audit with Kenya Water Towers Agency, says that in many places, people have recently moved into the forest illegally and started cutting it. Removing such people from water towers is necessary, according to Mwaniki with the Water Resources Authority, because the government has to look at the bigger picture to try to provide for all Kenyans.

Still, such actions are not without controversy. Population pressure *is* a major factor in deforestation. But Kenya also has a long history of government officials giving away public lands to cronies, as Maathai describes in her memoir, and that practice continues. And not everyone living in water towers is a recent arrival. In one case that came to international attention a few years ago, government officials evicted the Indigenous Sengwer people from their ancestral land in Embobut Forest, a water catchment in northwest Kenya. In fact, nature is healthier and biodiversity greater on the world's lands that Indigenous people manage or own, even compared with lands set aside for conservation by governments (as mentioned in chapter 1).

Musila told me more about how her agency is trying to navigate this sticky issue when we met at her office in a Nairobi business tower. She is a serious woman who earned her doctorate in forest ecology in Germany and conveys a sincere commitment to her work. "When I see a forest being cut, I feel really sad," she tells me.

"People don't understand what they are doing. Most people take a natural resource to be a free resource." Her agency is calculating the monetary value of the ecosystem services the water towers provide. By putting a price on these services, the agency can help people understand their worth and give them an incentive to conserve, she says. "Then you explain: by cutting this percentage of trees, this is the money you'll be using."

The agency gives relocated people payments for ecosystem services in the form of support for livelihoods projects, which also directly reduces pressure on forest land by addressing people's needs in a different way, Musila says. For example, the agency supports setting up people in beekeeping as a way to earn income, or growing fodder for their animals outside of the forest, or giving them a percentage of tourism fees.

*

Back in the Highlands, Kiarie, Ngure, and I visit a second restoration site, Zuti Forest, that has been recently replanted. It's also in Aberdare Forest Reserve. While part of Zuti Forest is protected, another part is being used for logging and grazing, despite it being earmarked for national park status. Enforcement is currently lacking, but it is slated to be closed off from these uses over time. According to one estimate, it may take a decade to fully transform into a national park.

Purity Muriithi, a young, pregnant ranger with Kenya Forest Service, greets us as we arrive. Wearing a camo jumper, boots, and a jaunty hat, carrying a machete, she points out recently planted saplings. Crabgrass covers the soil and cows graze—they've not yet been excluded from this area. Moving on, we walk under pencil cedars and past chest-high ferns and Mutundu trees with big, round leaves. Restoration began here in 2015 and is a cooperative project by the Green Belt Movement and Kenya Forest Service, with support from the water fund.

Ngure calls to me and points out a three-horned chameleon climbing a tree, a tiny modern-day triceratops. The lizard's colors are amazing: yellow, turquoise, and green knobbled skin. The eye facing me has radiating black lines and cranks around to check me out. When I get close, he opens his mouth in what appears to be a silent hiss, although it looks a little like he's laughing. I pick him up for a closer look—he's a bit smaller than my hand—and each foot's four toes splay in a Vulcan salute, gripping my fingers. I'm completely smitten, but Ngure can't believe it. He calls the others to see what this crazy *mzungu* is doing, and they are, to a person, equally startled. Muriithi, the ranger, shies away and vehemently exclaims, "I would never!"

I'm surprised; she's a ranger in a land with leopards and elephants, but she won't touch a small lizard? Later, in researching this aversion, I came across folk beliefs about the chameleon that are common across Africa. People believe it has poisonous spittle that will eat into your skin, or that its bite will never heal, or turn you infertile, or switch your sex. Because it can change colors, people believe witch doctors can use it to send bad spirits your way. According to science, chameleons are harmless to people. A Kenyan academic told me that she used to see them everywhere when she was growing up, in classrooms, bedrooms, playgrounds. But now she can only find them in natural forests.

I return my colorful friend to the tree and we humans continue walking downhill. Ducking under an electric fence and into an area replanted earlier, where the mixed indigenous trees are growing thick and lush, we can see nearly down to Gikira River, which Kiarie says "comes straight from the Aberdares."

Seedlings for restoration projects like this one are supplied by a Kenya Forest Service nursery and also by small family farms. The projects often employ local women to do the planting. The Nature Conservancy says more than two thousand women have earned income, wood, and fruit, while planting more than three million trees throughout the Aberdare highlands.

Planting a Trillion Trees: Beware

As the world has gotten more serious about mitigating climate change, massive-scale tree planting has caught the political winds worldwide. In 2020, the World Economic Forum in Davos, Switzerland, called for the world to plant a trillion trees to help store carbon dioxide. The United Nations Environment Programme has a Billion Tree Campaign. Governments jumped on this seemingly simple solution: the European Union promised three billion trees. The African Union pledged to bring about 386,000 square miles across the continent under restoration by 2030. China was ahead of the curve: its Great Green Wall aiming to halt desertification has planted more than sixty-six billion trees since 1978.

Planting trees is a gut-level environmental good: Who doesn't love trees? It's great that people are beginning to recognize that nature-based solutions need to be part of the strategy to reduce climate change. But like everything to do with the natural world, a singular focus can spawn multiple unintended consequences. Trees are not just carbon sticks. They offer other equally important benefits: water security, critical habitat for critters and plants, shade and cooling, food and materials for local people's needs. At the same time, tree planting may cause problems if not well thought-out.

Attempts to use trees for carbon reductions—often through accounting schemes selling carbon credits or offsets—have led to missteps such as single-species plantations of fast-growing trees. These projects sought to absorb as much carbon as possible and, in the process, financed a big loss for biodiversity. In many cases, the species planted were nonnative, such as acacia, *Eucalyptus*, and pine. Monoculture plantations are ecologically weak, prone to mass die-offs from disease, forest fires, drought, and even weather. It's akin to how animals raised in factory farms are more susceptible to disease.

I ask Leisher about this tendency to plant single species. "Nature's saving grace is that it . . . has all these interconnections; when one is broken, another moves into place. When you start simplifying, it becomes much more fragile," he warns. He is evoking the same principles of systems complexity that Canadian forest researcher Suzanne Simard discovered with the mycelium and mycorrhizal networks that allow trees to share information, water, and food (as mentioned in chapter 3). Near where I live in Victoria, BC, stumps cut by loggers decades ago live on—despite having no leaves to photosynthesize food—because neighboring trees are sharing resources.

Intact, healthy forests are also better at storing carbon, sequestering twice as much as a planted monoculture, according to the authors of the Global Deal for Nature conservation plan. In part that's because carbon is stored in the soil too, and installing a plantation disturbs and releases that cache.

*

Before planting trees willy nilly, we also need to consider the massive role they play in the water cycle. Trees generate rain. Scientists have long known that evaporation from oceans and lakes moves water into the air that later falls as rain. They have also understood that trees, like other plants, take up water from their roots and exhale it through their leaves into the atmosphere, where it can form precipitation. But they used to think that trees were small contributors to rain. Then in 1979, Brazilian meteorologist Eneas Salati, using the fact that water from different sources has different chemical signatures, showed that half the rain in the Amazon rainforest came from the trees themselves. Trees can also generate rain far away: the Amazon produces precipitation as far away as Texas; the Congo forest waters the US Midwest; forests in Southeast Asia influence rain in the Balkans. The flip side—as Kenyan farmers have witnessed firsthand and which has been borne out in studies—is that deforestation can reduce rainfall magnitude.

Trees may also pull rain toward themselves. An intriguing theory argues that trees exhaling water vapor also drive wind to carry water into the interior of continents (rather than the wind driving the water). Russian physicist Anastassia Makarieva and her mentor Victor Gorshkov came up with this theory, which they call the biotic pump. They argue that the interior of continents—South America's Amazon forest, Russia's taiga, Canada's boreal—get rainfall thanks to an expansive network of trees. It might also explain why a deforested continent like Australia has an arid interior.

Some people who are thinking about the relationship between trees and water have come to the incorrect conclusion that trees are

bad because they drink a lot of water and therefore reduce local water supply. But the idea that trees just suck up water is based on a
2005 study in *Science* that's held way too much sway, says Douglas
Sheil, a forest ecologist and professor at Wageningen University &
Research in the Netherlands. The researchers looked at twenty-six
long-term studies of tree planting and found significant decreases in
annual stream flow. But almost all those studies were looking at *Eucalyptus* or pine plantations. The right trees—or plants—in the right
place are less likely to deplete local water, Sheil tells me via phone.

In Kenya, just as in Peru and California, people have learned the
lesson about the wrong trees the hard way. Public sentiment has now
turned against *Eucalyptus*, native to Australia, as people have seen
how much water they drink. In Kenya, the Nature Conservancy is
working with farmers to replace *Eucalyptus*, which people grow for
wood, with bamboo, a plant that uses less water, holds the soil, and
can be harvested repeatedly.

Similarly, people have found that planting trees in landscapes
where they don't belong—such as in native grasslands like Ecuador's *páramo*—can also reduce water supplies in local streams. Plus,
if people are planting trees in grasslands to store carbon, that's dou-
bly misguided because healthy grasslands store carbon too. It also
takes living space away from plants and animals who prefer grassland to forest.

In California's northern mountains, some people wanted to cut
native trees, fearing that they were drinking scarce water supplies.
But just like other single-minded "problem solving" in natural ecosystems, the idea of cutting trees to save water turns out to be erro-
neous. In a natural experiment, trees are dying en masse across the
western United States from pine beetle outbreaks (a consequence of
the warming climate that does not kill enough native beetle larvae in
the winter). But stream flows are not increasing as trees stop drinking the water. A survey of beetle-killed forests in Colorado found
that, instead, the loss of so many trees has led to sunnier, warmer,
windier zones where winter snow and summer rain evaporate rather
than slowly recharging groundwater.

To summarize: it's complicated. But it's also not so complicated, says Leisher. Just restore indigenous ecosystems—or give them space to restore themselves. It's beginning to look like a hands-off approach might work best. Replanting indigenous plants, while generating local jobs, may not achieve the best results for carbon storage or for water.

"The problem with people who take money for this is they feel they have to do something," says Sheil. "There's a big obsession with planting trees because then you can show what you're doing. But one thing about trees is, they're pretty good about planting themselves." The first step is to withdraw human activities that are killing the trees in the first place, Sheil emphasizes. But then it's often sufficient to allow the forest and native plants to seed and recolonize themselves. Planting is only required in specific cases as a way to kick-start a self-sustaining process. Money collected to restore trees, grasslands, wetlands, or other native ecosystems could be put toward buying and conserving land, enforcing protection, removing invasive species, or supporting livelihood or fuel alternatives for people who were denuding the land.

Water Returns

Fortunately, in Kenya, many people seem to understand these ecological principles and the connections throughout the watershed, from the top of the water tower in Aberdare National Park's moorlands, to indigenous forest restoration on the mountain flanks below, to sustainable farming practices, on down to the human residents of Nairobi and beyond.

More broadly, Kenya's various avant-garde approaches to holistic water management could mean a radical change from the usual tendencies, when facing scarcity, to grab more water from somewhere else. That's not to say that some people in Kenya don't want more dams as a hedge against drought. But Kihara, the Nature Conservancy director, cautions, "Dams don't magically bring more water." They only store water that is already there.

Also, dams cost a lot of money—money Kenya often doesn't have and cannot borrow, according to Mwaniki at the Water Resources

Authority. Instead, his agency and others are trying to conserve watersheds. "We rely on available surface water or groundwater," he says. "That's why we harvest water and store it." Kenya's water policies "take into consideration upstream producers and downstream water users."

Talk of harvesting water, or calling upstream people "water producers," is analogous to the notion of farmers in the Andes Mountains "planting" water in Peru. This language might please sociohydrologists, as it clearly communicates the role that human actions have on water supply. Foresters, in replanting trees, produce water, as evidenced by those seventeen returned streams the Nature Conservancy has counted in the Sagana–Gura watershed. Farmers produce water too as they grow other crops. Some have even brought back flow to desiccated streams on their own land.

I have the chance to see a stream reborn with agricultural officer Kiarie. In the Sagana–Gura watershed near Mukurweini town, we visit Josephine Wanjiku Mwangi, a grandmother dressed in an orange-and-blue *khanga* wrapped as a skirt, a white rolled head wrap, and orange plastic sandals.

She welcomes us to her farm and ushers us past her cow eating grass from a manger and a black-and-white cat lazing about. Below, Josephine and her husband Joseph Mwangi grow bananas, coffee, and avocados. She shows us the numerous bench terraces they have created over the last year. First, they dig a trench horizontally across the hillside, which captures rain as it flows downslope along with the soil it carries. Eventually, soil deposition creates a flat, horizontal bench. About ten feet down the hill, they create the next bench, so that in the end they have a series of steps rather than a steep grade. By slowing the water and giving it time to infiltrate the soil, crops can then use that rain.

Downhill, Joseph is digging another trench by hand with a pickax. Hauling the tool overhead, he brings it down with force into the red earth. Kiarie tells me she can't believe how many terraces this couple has built since she last visited.

The hard work is paying off, Josephine tells us. The terraces can be used to grow arrowroot, sometimes called coco yam. This culturally

FIGURE 8.2 Arrowroot, or coco yam, is a culturally important food in the Highlands of Kenya. The crop needs a lot of water, so people tend to grow it along streams, but this causes erosion, evident in the cloudy water. An alternative is for farmers to dig terraces that can capture rain, providing the coco yams with the water they need.
Photo © Erica Gies

important food is often grown right in streams or alongside because it needs a lot of water—but that practice increases erosion. Growing it in the terraces works, says Josephine, quite pleased as she shows us a stocky white tuber coated in red dirt.

FIGURE 8.3 Josephine Wanjiku Mwangi uses her machete to hold back napier grass. She is showing me a seeping pool of water among the roots. Her spring has returned, thanks to the terraces she and her husband dug on their steep farm to retain rainfall.
Photo © Erica Gies

Before building the terraces, the Mwangis were barely able to grow crops, she tells us. Now they are reaping good harvests—and their spring has returned! To see this small miracle, we walk for fifteen minutes downhill on the Mwangis' one-and-a-half acres, sliding on banana leaves laid atop red mud gluey from the previous night's rain. Finally, Josephine stops, bends over, and separates the tall napier grass with her machete. I lean down and see light reflected off the seeping water. It's a subtle but important sign that this plot of land, its crops, its people and other animals have enough water and are thriving.

· 9 ·

Sedimental Journey

WHERE FRESH WATER MEETS SALT

Downstream from Kenya's towers, some of the fresh water flowing off the mountains and across its plateaus eventually finds its way to the coast. There it pushes and twines and gives in endless negotiation with the sea. That waltz is ancient, but now two-fifths of humanity—billions of people, including me—live within about sixty miles of the world's coastlines, in the midst of that dynamism. And we've made those dynamics more extreme with the ways we're changing land, rivers, and climate. It's becoming obvious to us coastal residents that our homes are threatened by increasingly frequent flooding from overland, from beneath, and from the sea. Creeping sea levels have become a rallying cry at climate protests such as Fridays for Future inspired by Swedish teenager Greta Thunberg and Extinction Rebellion die-ins. Placards and protesters shout, "Seas are rising, and so are we!"

Back where I'm from, in the San Francisco Bay Area, people are fighting these consequences of human actions by wielding nature. Humans are often inclined to build seawalls to protect coastal communities from encroaching oceans, but those require constant, expensive maintenance. Natural coastal ecosystems, on the other hand—tidal marshes, barrier islands, coral reefs, seagrass beds, dunes, gravel beaches, kelp and mangrove forests—have the ability to sustain themselves. And if left intact, they slow fresh and tidal water, acting as a buffer, providing flexible and resilient protection for human communities. They can reduce by half the number of lives and properties at risk from storm surges and sea-level rise, according to a study in *Nature Climate Change*. To learn more about this opportunity, I met up with some coastal water detectives on the east side of San Francisco Bay, just south of Hayward.

It's a golden summer day in 2018, and we're standing on a low coastal levee, overlooking a pond at Eden Landing Ecological Reserve that looks positively apocalyptic. Algae paint ruddy swirls in the brown water, its edge crusted hard with sparkling salt. As a breeze eases off the bay, a squadron of pelicans sails by, en route to more appetizing hunting grounds.

This pond is a legacy of a salt industry that has moved elsewhere. A few decades ago, when flying into San Francisco or San Jose, the ground beneath looked like a giant's Easter egg dip. Ponds of blue, yellow, green, red, purple, orange, and pink ringed the South Bay. People had built low levees in semicircles from the shore, sectioning off portions of the bay to let the water evaporate, leaving behind the salt. The different colors were caused by varying levels of salinity and the types of organisms who could live in them—algae, bacteria, brine shrimp.

But today these former industrial sites present opportunity. Reverting more of this coastline from salt ponds and flood-control levees back to natural ecosystems could help protect the San Francisco Bay Area from sea-level rise. The pond I'm standing by is awaiting change. But as I turn my gaze west, toward the Golden Gate, I see something that looks much more natural: Whale Tail Marsh, a low-lying pastiche of variegated greens and tawny yellows whose restoration began twenty years ago. These two areas, side by side, are symbolic of nature's resiliency, where we allow it, and how much work remains. Now, as climate change accelerates, coastal restorationists are running out of time.

*

Sea levels are rising because the greenhouse gases we are emitting are warming the atmosphere, which is melting glaciers and polar ice shelves, and their accumulated water is swelling oceans. Also driving sea-level rise is the fact that, as oceans warm, water expands. Coastal ecosystems help not just with the impacts of climate change by buffering human communities from the rising seas; as we've seen with freshwater wetlands, they also help slow its progress by storing carbon dioxide as they grow. This potent repository has been dubbed "blue carbon." Coastal ecosystems such as salt marshes, seagrass meadows, and mangroves sequester up to three times more carbon per acre than tropical forests. That's because the water flooding them all or part of the time reduces available oxygen, so not all the carbon captured by their plants is metabolized by microorganisms and released when they die and decompose. And sea-level rise may even increase their carbon storage—if they don't drown too quickly. Researchers looked at 345 wetland sites on six continents over a six-thousand-year record. They found that when faced with sea-level rise, coastal wetlands store two to nine times more carbon dioxide in their soils.

Also similar to freshwater systems, these ecosystems process and break down nutrients and pollutants, including sulfur, phosphorus, and nitrogen. Yet unlike forests, the role that wetlands protection can play in slowing climate change is barely mentioned in global climate change agreements.

Despite these benefits of coastal ecosystems being broadly understood, people continue to destroy them. Over the past half century, 50 percent of salt marshes, 35 percent of mangroves, 30 percent of coral reefs, and 85 percent of shellfish reefs worldwide have been destroyed, seriously undermining coastal resilience. According to one recent estimate, about 1,313–3,784 square miles of these coastal ecosystems are destroyed each year, releasing their stored carbon into the atmosphere. In their place, developers erect buildings, sell them, and move on; but those homes and businesses are likely not long for this world due to sea-level rise. Cities allow such ill-planned development because they reap property taxes in the short term. But in so doing, they fail to account for the astronomical future costs to protect the properties and to continue providing services such as water, electricity, and roads.

In San Francisco Bay, local scientists, governments, and NGOs have a head start. For decades they've been working to restore some of the more than 90 percent of San Francisco Bay wetlands lost to development. Initially, their motivations included reducing flooding; giving local people recreational access to the shoreline; and restoring habitat for endangered species such as the Ridgway's rail, who nests on the marsh, and the pickleweed-loving salt marsh harvest mouse. These efforts have already made San Francisco Bay one of the world's largest wetlands restoration projects. But as climate change accelerates and the threat of sea-level rise has sharpened into focus, people beyond the restoration community have begun to realize that the tidal marshes could be a first line of defense, cushioning human infrastructure. Now restorationists here have made it their explicit goal to reestablish as much marsh as possible to help protect against sea-level rise before it's too late.

Related efforts are getting underway worldwide, such as repairing Southeast Asia's devastated mangrove forests. Cut for charcoal and to make way for shrimp farms, their absence leaves many coastal areas unprotected. In Vietnam, the low-lying Mekong Delta produces half of that country's staple crop, rice. Today, salt water is moving into the delta at an unprecedented rate, ruining cropland. That's due to the lost mangroves, sea-level rise, and upstream dams on the Mekong River that hold back critical fresh water and sediment. Now Vietnam is starting to adapt, by diversifying crops and restoring mangroves. Across the globe, Slow Water practitioners, Coastal Edition, are recognizing that every coast is unique, but all face common challenges: rising seas; increasing temperatures; bigger, slower storms; and development along the ocean's edge.

Risk and More Risk

Sea levels have risen an average of six to eight inches over the last century, and the pace is accelerating—half that rise has come since 1993. That makes hurricanes, typhoons, and rain torrents more devastating for coastal communities. As these storms cross warmer oceans, they gather more energy and water. They are also traveling about 10 percent more slowly now than in the middle of last century, meaning their winds and rains linger longer. That was true of Hurricane Dorian, which sat over the Bahamas for two days in 2019, dumping thirty-six inches of rain and at times moving just one mile per hour.

Devastating floods have become ubiquitous in recent years. Summer and fall 2021 saw deadly flooding in many places, including India, Thailand, the Philippines, and US states such as Alabama, New York, and New Jersey. In January 2020, Indonesia's capital Jakarta flooded to five feet after a big rainstorm, killing sixty-six people. The country's president then announced the capital, with a population of thirty million, will move to higher elevation on the island of Borneo. Sea-level rise also pushes up hurricane storm surges that flood further inland. Tropical Cyclone Idai in 2019 killed at least 1,300 people and caused catastrophic damage in Mozambique, Zimbabwe, and Malawi.

Coastal cities are not just flooding when stormy but also during high king tides, sometimes called sunny-day flooding. Since 2000, the frequency of sunny-day flooding has increased by five times in some US cities, damaging homes, swamping roads, and tainting drinking water, according to a 2020 report from the US National Oceanic and Atmospheric Administration (NOAA). Eagle Point, Texas, on the Gulf Coast experienced high-tide flooding sixty-four days in 2019. Cities including Galveston, Corpus Christi, and Morgans Point, Texas; Annapolis, Maryland; and Charleston, South Carolina saw from thirteen to twenty-two days of flooding each. NOAA predicted that by 2050, the median period of high-tide flooding on US coastlines could jump from seven to fifteen days to between twenty-five and seventy-five.

These floods don't even need to inundate a neighborhood to cause trouble. They can prevent people from reaching or using their homes by swamping critical infrastructure like roads or sewage treatment plants, many of which lie on coasts. They can obliterate beaches that coastal cities rely on to attract tourists. Seawater can infiltrate underground and move inland, turning drinking water salty, making land too saline to grow crops, or pushing up fresh ground-water to flood homes from below. Saltwater intrusion is already a significant problem not just in Vietnam's Mekong Delta but in many other places, including parts of Florida; Basra, Iraq; Bangkok, Thailand; and Tripoli, Libya.

When people respond to these threats by building a wall, they often reap unintended consequences. Seawalls don't protect against salt water moving in underground. They are also brittle. Hard barriers lock the border between land and sea. Trying to force stasis on a mobile system ultimately creates the opposite effect, leading to further erosion and requiring ongoing maintenance. Waves scour the ground at the bottom of the seawall, digging out a hole and ultimately undermining it. Plus, the wall blocks sand and sediment from coming back in to replenish beaches. Like levees, seawalls are neighbor-hostile, deflecting wave energy away from one area to a nearby, unprotected area, eroding that land faster. Because poor and otherwise marginalized people often live in low-lying areas and lack funds to build their own seawalls, they bear the brunt when neighbors armor up.

Like dams, seawalls are expensive and, depending on their size, can take a long time to build—in some cases, making them ineffectual by the time they are completed. The US Army Corps of Engineers recently finished a $14 billion levee network in New Orleans that is already sinking due to subsidence and is expected to be ineffective by 2023. Both Boston and New York City have considered and rejected giant flood barriers—mostly because they won't work. Boston weighed a 3.8-mile-long seawall that would have cost nearly $12 billion, taken thirty years to build, and caused major environmental damage. Instead, it opted to add 67 acres of green space along the water, restore 122 acres of tidal marsh, and elevate some flood-prone areas.

Fortresses of Mud

Unlike seawalls, tidal marshes have a superpower against sea-level rise. It's not just that they are a buffer between the water and human infrastructure, sapping energy from storm surges and blocking the highest tides. Marshes can actually grow vertically, keeping pace with sea-level rise by trapping sediment in their vegetation, which decomposes and then regrows. To perform this trick, they need three ingredients: sediment, space, and time.

"Marshes are in a dynamic equilibrium with the water level. It's been clearly shown that, even at pretty high rates of sea-level rise, if there's enough suspended sediment, they can keep pace," John Bourgeois tells me that day at Eden Landing. A laid-back guy originally from Louisiana, Bourgeois has become a significant figure in San Francisco Bay's marsh restoration. For nine years he was executive manager of the South Bay Salt Pond Restoration Project, a public–private partnership that manages wetlands restoration of former salt ponds in Eden Landing and other sites in the South Bay.

Restorers have documented the speed of marsh growth. In the town of Alviso, near San Jose on the edge of the South Bay, a protective marsh accreted more than six feet in twenty-five years. Another long-term restorationist and water detective, Letitia Grenier, interviewed local people about this feat. A bright, warm woman who's been studying the ecology of the bay for two decades, she is now director of the Resilient Landscapes program at the San Francisco Estuary Institute (SFEI), a scientific research organization that studies water, wetlands, wildlife, and landscapes. Alviso residents told Grenier that where they used to dock their boats along the slough with plenty of draft, there is now marsh. "When natural processes deliver the sediment, the plants grow up through it, and the living marsh stays on top and the sediment keeps accreting," she tells me when I visit her office in Richmond, California.

Can San Francisco Bay's marshes even keep up with fast sea-level rise? I ask.

"The answer is yes, if they have enough sediment. Although," she amends, "when sea-level rise accelerates very rapidly, it's anybody's guess, because we've never seen that before."

Restorationists in San Francisco Bay may have the time they need to restore all available coastal land because sea-level rise is happening a bit more slowly on the US West Coast than on the East and Gulf

Coasts for complex reasons. California has seen about six inches of sea-level rise over the last century, although that's still enough for some low spots to start flooding during king tides, such as San Francisco's famous Embarcadero at the foot of Market Street and parts of San Mateo and Marin Counties.

But there's not a lot of time. The rate of sea-level rise is expected to accelerate as the century goes on. Levels could rise by as much as seven feet by 2100, the California Ocean Protection Council estimates, and that would threaten infrastructure, including airports, electricity plants, transportation, and drinking-water facilities— many of which lie low and close to the bay.

Bourgeois and Grenier are part of a vast, loose group of scientists and local government officials who have been working to fulfill a goal set in 1999 to restore more than half the 190,000 acres of marshes the bay had historically. Although precise documentation is lacking, experts estimate that, at the low point in the mid-twentieth century, just 10 percent of the historic marshes remained. Today, about 28 percent of the historic marsh area, or approximately 53,000 acres, is marsh or being restored to marsh. In 2016, marsh allies refined their goal, making it an explicit objective to restore as much acreage as possible by 2030 for maximum resilience to sea-level rise. Currently planned restoration projects would add at least another 22,000 acres of marsh for a total of almost 40 percent of the original marsh area. The easiest large tracts of land to acquire and restore have already been slated for repair. Remaining open space around the bay is plentiful in certain areas but is privately owned, and some has subsided below sea level, which makes it expensive and difficult to restore to tidal elevations. Still, the goal is 100,000 acres total, if the group can leverage other opportunities for restoration.

Getting that space, someway, somehow, is one of the critical three ingredients marshes need to work their wonders. But in the heavily populated Bay Area, like in many other water-fronting places around the world, space is at a premium. Just behind the marshes lie freeways, sewage treatment plants, technology campuses such as Facebook and Google, and local neighborhoods. That human infrastructure is problematic because marshes need room to migrate inland and because they are part of larger systems.

"You can't just think of the marsh," Grenier tells me—another systems theory lesson the scientists have learned by doing. When they started restoring Bay Area tidal marshes, they thought, "We'll just breach [the levee creating] this salt pond that used to be a marsh," she tells me. "We'll let the sediment come in, it'll be a marsh again. Boom. We're done." As they studied how marshes work, they realized the truth: "These are much bigger systems." Thanks to fresh and salt water moving through them, marshes interact with several other neighboring, intertwined ecosystems, both upstream and out into the deeper water of the bay. It's a mobile spectrum: the subtidal ecosystem, populated by eelgrass and oysters; mudflats where egrets and endangered California least terns hunt; the low marsh with its Pacific cordgrass; the marsh plain featuring pickleweed, a tiny segmented succulent that—surprise—tastes salty; the high marsh with its gum plant; and finally, the transition zone reaching into the uplands, home to shrubs and multiple species of oaks. When free from human-made barriers, the zones exchange critical sediment and nutrients with the help of streams flowing down from the uplands and tides pushing inland.

"If you just have the marsh, it's not as resilient as if you have the full system because each element protects what's behind it," Grenier explains. In their natural state, these ecosystems achieve "dynamic stability"—not an oxymoron, but a resilient state of flux. They are self-maintaining. That's why marshes need space: they grow upward in part by slowly marching inland. Where they have room to move, they can creep over sediment in the uplands, feeding themselves as they go.

Marshes aren't the only borderline between land and water along San Francisco Bay. Historically there were about twenty-seven miles of sand, shell hash, gravel, and cobble beaches along the bay. Now there are hardly any; they've been replaced by artificial fill, riprap, or walls of large stones or concrete. But like the tidal marshes, the gravel of days gone by also absorbed wave power, like a football player who rolls when tackled to diffuse force on the body. Because riprap and seawalls are hard barriers, they don't absorb wave energy but instead push back against it, banking it onto neighboring marshes, eroding them. Other restoration projects around the bay are now seeking to replace gravel, sand, or shell beaches in some places.

To find space for these natural cushions, water detectives look for opportunity from the human side, restoring places that have become available because they are now less attractive to people, such as the salt ponds. But they are also looking to water to guide them. Like the landscape architects at Turenscape in Beijing, or KK's sleuthing in Chennai, historical ecologists at SFEI seek to discover what water once did in this landscape before development obscured its patterns. They then create maps showing where water wants to go, overlaid on modern development, that can help people understand where keeping water out may be a losing battle, and where relinquishing acres to nature can have a big impact on protecting adjacent human habitat.

Forensic Ecology

On a rainy January day in 2019, I meet up with two of SFEI's historical ecologists, Sean Baumgarten and Erin Beller (now an ecology program manager for Google), at the Bancroft Library at the University of California, Berkeley, so they can show me the ropes. As is typical with collections of rare books and historical documents, the library's treasures can only be perused in the reading room, which we enter through multiple checkpoints after agreeing to many idiosyncratic rules. It's all in a day's work for Baumgarten and Beller, who spend months digging through troves of local historical societies' documents, city and county agency records, and state and national archives. Maps, landscape and aerial photographs, agricultural surveys of soil types, legal testimonies, travelogues, and scientific studies of tree cores and pollen all provide glimpses of lost landscapes and, crucially, evidence of where water once flowed and slowed.

Prior to our visit, Beller had created a list of hunting targets, such as now-obsolete names of archaic tiny local creeks or hills that disappeared into history when they were cut down and covered over by development. Using those names as search terms, they plumbed archives and catalogs, putting in advance requests to the librarians to retrieve old documents from the library's gold mine of rare information. Together we dug into the resulting pile.

The first book I read is the traveler's account of San Francisco mentioned in chapter 3, written a few years ahead of the Gold Rush of the mid-1800s. The author, William H. Thomes, describes San Francisco Bay: "At high water we could clear the mud flats that ran out some twenty fathoms [120 feet] from the beach, and mean, sticky mud it was, extending downward for many feet, as builders afterward discovered." Mudflats are part of the interconnected systems Grenier was telling me about—important habitat and sediment supply on the bay side of marshes.

Thomes mentions a hazard to ships called Blossom Rock that is no longer: "The United States made short work of that rock, after taking possession of the country. A few hundred pounds of powder, or some other substance, removed the stumbling block to navigation, and ships are no longer required to give the point of Telegraph Hill a wide berth." Wildlife at that time was prodigious. Thomes describes islands in the bay "covered with sea fowl and seals," and writes of encountering rattlesnakes and grizzly bears while walking across what is now San Francisco.

Next up on my desk at the library is a handwritten legal document from the 1860s about a land rights dispute. The sheaf of papers is tied at the top, and I struggle to decipher the slanted cursive inside. Beller, on the other hand, quickly homes in on some pertinent sections, including testimonies from people who lived around Mission Dolores—the church founded by the Franciscan order in 1776 in the heart of modern San Francisco's Mission District. One person's account mentions proximity to a river and well. Another witness drew a simple map showing trees, which Beller reads as a reliable sign of water in the chaparral landscape.

Beller tells me of an account she found from 1843 of someone living in an adobe house at Montgomery and Market Streets: "There was no fresh water for miles around." From that standpoint, "the house was a pain in the butt." Tidbits like that "help animate the landscape and promote a visceral sense of what's it's like to live in that place," she says. And although a drawing of trees or a passing mention of mudflats are not much to go on, each piece of information is a snapshot in time and space. "In the aggregate, [these pieces of information] help to figure out surface water patterns in the cities," she explains, or the extent of mudflats and marshes ringing the coast.

Researchers rate each datum for accuracy using a system developed by SFEI. Key questions include: What is it? How big is it? Where is it? "A lot of it is context," Beller explains. "Who made this? What was their motivation? What would their blind spots have potentially been?" An advertisement exhorting people from New Jersey to move to California would get a lower credibility rating than a scientific survey. Over its more than twenty years of work, SFEI has come to know and trust certain sources—and even individual mapmakers who were especially persnickety about details.

The data are raw ingredients that inform visualizations of landscapes past. Environmental scientists and analysts at SFEI weave together the fragments of information by creating a geographic information system map, just as Turenscape's landscape designers do in China. The key, according to Beller, is to "evaluate sources in a way that's transparent and replicable for other people."

SFEI often starts with nineteenth-century US Geographic Survey topographic maps and hydrologic surveys, coastal navigation maps, historical and modern aerial photographs, and century-old maps from the US Department of Agriculture demarcating thirty different types of soil. (The latter is a good indicator of water capacity and vegetation and habitat type.) Then the historical ecologists add their archival finds, weighted to reflect their confidence in each bit. Their approach has been taken up and adapted by others in this small field in different regions. Fortified with these detailed maps of what once was, detectives can understand what water wants in a particular place and look for opportunities to accommodate it.

Searching for Sediment

The regionwide effort by scientists and local governments to restore as much tidal marsh as possible is well underway. But in their ambition, they are coming up against a challenge. They need more of the third and final ingredient that marshes require to heal and grow: sediment. Existing tidal marshes and mudflats are unlikely to receive enough sediment naturally in order to survive sea-level rise this century, according to a recent report from SFEI. Without supplemental sediment, many will drown by 2100.

In a natural system, fresh water from upstream rivers and brackish and salt water moved by the tides play critical roles in delivering sediment to marshes. But humans are blocking sediment delivery. This phenomenon affects San Francisco Bay, as well as many other river deltas around the world: the Mississippi, the Niger in Nigeria, the Indus in Pakistan, the Ganges–Meghna–Brahmaputra in India and Bangladesh, the Irrawaddy in Myanmar, the Mekong in Vietnam. Upstream dams trap sediment behind them. And leveed, channelized rivers shoot water and the reduced sediment it carries off the land and out beyond tidal mudflats. Depriving the near shore of sediment leaves a delta without enough raw material to keep up with normal coastal erosion, let alone rising seas. That's one reason Louisiana has been losing land. And hardening the shoreline blocks tide-carried sediment from building land seaside.

If the San Francisco Bay restorationists are to be able to fulfill their plans to revive more of the marshes, they will need extra sediment. The old salt ponds are one to two feet below sea level for a few reasons. The biggest: diking them off from the bay blocked tides bearing land-building sediment. Groundwater pumping also caused the land to sink. Dave Halsing, current manager of the South Bay Salt Pond Restoration Project, tells me on the phone that just opening a pond to the tides and allowing it to accrete naturally might take five to twenty years depending on the depth of the pond. That could be too slow to beat sea-level rise. The need for sediment has restorationists searching high and low for more so they can fill the subsided holes to jumpstart the natural restoration process. But they require a lot.

The lead author of SFEI's sediment report, geomorphologist Scott Dusterhoff, quantified the sediment required both to maintain existing marshes and mudflats and for additional planned marsh restoration. If sea levels rise by 6.9 feet by 2100—a moderate-to-high-risk scenario in which humanity doesn't find the will to reduce emissions—they'll need more than six hundred million tons of sediment.

Some could come from upstream. Dusterhoff calculated that by 2100, without intervention, creeks could deliver 176 million to about 309 million tons of sediment to the bay. The variation is due to drier or wetter future scenarios. With less rain, there will be less water flowing down creeks to the bay carrying less sediment. With some intervention, creeks could deliver more. According to Dusterhoff, an additional 353 million tons of sediment could be harvested by 2100 from bay dredging for ship traffic. And another 165 million tons excavated during local construction could be put to use. Added up, there's likely enough sediment in the region to make up the deficit. It's a matter of getting the dirt to where it's needed.

*

Until recently, San Francisco Bay has had enough sediment—sadly, due to an environmental catastrophe perpetrated in part by my ancestors. One branch of my mother's family moved to California from Maine in the mid-1800s, lured, like so many others, by gold. Arriving in San Francisco, they made their way north to the foothills of the Sacramento Valley, where tributaries coming off the Sierra Nevada feed into the Sacramento River and, ultimately, San Francisco Bay. All the big nuggets were picked up in the first few years, so miners turned to other methods. Romantic lore in California heralds the independent stakeholder with a pickax and pan, squatting alongside the river beside a mule, patiently screening gold from sediment. But in 1853, people began hydraulic mining, the practice of training high-powered hoses on river hillsides to reveal veins of gold, tearing out plants and sloughing soil in the process.

My great-great-grandfather, great-great-uncle, and great-grandfather did this work along the Feather River near the Sacramento Valley town of Oroville. At the peak, hydraulic miners blew 7.2 million tons of sediment a year down rivers toward the bay. It's an industrial abuse that remains visible today in vertical cliffs of red and white dirt, still naked of trees. Over a period of about thirty years, so much dirt was carried downstream that it raised the Sacramento River's bed by five to thirty feet and settled in a thick layer on the bottom of San Francisco Bay. Finally, in 1884, a US District Court, spurred into action by a farmer-led opposition group called the Anti-Debris Association, banned the practice.

The unnatural load of sediment from hydraulic mining was about eight times the magnitude of natural flows. But after 150 years, that pulse is settling out, and now sediment scarcity looms in the bay.

In later decades, another type of sediment dump threatened the bay: people were filling it in to create new land for development. The shallow mudflats and tidal marshes were easy targets. By the 1950s, when my mom was growing up on the peninsula south of San Francisco, all but 10 percent of the bay's marshes had been diked off or filled in for human uses such as agriculture, garbage dumps, sewage treatment plants, navy bases, airports, and salt ponds, diminishing the size of the bay by one-third. My grandparents had moved from the Sacramento Valley to the Bay Area for college and ultimately settled off Marsh Road, a name whose significance—that the eastern portion was built atop a marsh—largely escaped me until I began this reporting.

On another summer day in 2018, I returned to this place from my childhood. Driving down Highway 101, I took the Marsh Road exit to visit the Ravenswood/Bedwell restoration site with Bourgeois. We climb a small hill while he outlines the project so far. It was originally marsh, then salt pond, then dump. Now it's a park, capped with dirt and replanted with native plants.

"Which part is the dump?" I ask.

"We're standing on it," he replies.

"OK, so historically there wasn't a rise of land here?"

"No," he explains. "We're standing on a pile of garbage."

In fact, he tells me, the South Bay has no natural highlands along the shore except for Coyote Hills. He gestures toward them, across the bay. Other than that, "any high points on the edge of the bay are garbage," he says. "Especially if you go down to Alviso: there are lots of hills and it's all garbage."

Strangely enough, this particular dump was one of my mom's favorite places during her childhood. She used to visit it on weekend outings with my grandfather. After pruning the peach and walnut trees he'd planted in his suburban yard, "he would fill an old trailer we had and hitch it to the back of the olive-green Packard, and off we would go," she tells me. She and her sister both sat on the front bench seat with him. They drove out Marsh Road and across Bayshore, the old name for 101 because it was built on the original shoreline of the bay. Land to the east of 101 is largely fill. After paying the dumping fee at a small shack, "we would drive down a dirt road, through smoke and slowly burning piles of trash." Thousands of seagulls were flying around, very noisy, scavenging what they could. "It was great fun," she reminisces. "Part of it was being together and sharing the experience, part of it was watching the fires burn, and part of it was a simple pleasure in a simpler time."

But not long after, public attitudes shifted, as people began to realize the damage being done to the bay. The breaking point was a 1959 proposal from the Army Corps of Engineers to fill 60 percent of the remaining water in the bay for more buildings, which would have left just a narrow passageway for ships in one of the world's great natural harbors. Public sentiment turned to outrage. Three local women, Sylvia McLaughlin, Kay Kerr, and Esther Gulick, founded an organization the next year called Save the Bay, whose advocacy ultimately led to new laws and government agencies to protect the bay. In the 1970s, local scientists began some of the world's first marsh restoration projects.

Free Alameda Creek

Those founding women would likely be surprised at how much has been accomplished since then and the ongoing ambition of people working to restore the bay today. One of them is Gena Morgis, a landscape designer with the New York landscape architecture firm Scape who visited Eden Landing with me as part of another Dutch–local partnership called Resilient by Design. The goal was to come up with proactive concepts for natural protection around the bay that would anticipate and prevent disasters. The Scape team was looking upstream for ways to free the natural supply of sediment from rivers and creeks blocked by dams. In their project, dubbed Public Sediment, they focused on Alameda Creek. As the largest watershed feeding San Francisco Bay, it has the most sediment potential. Before it was heavily engineered during the twentieth century, the creek used to deliver its dirt to the marshes at Eden Landing.

Morgis is one of the water detectives present on my visit to that apocalyptic pond at Eden Landing and its restored neighbor. A young woman dressed more for East Coast office than West Coast field, Morgis bubbles over with an endearingly nerdy enthusiasm for sediment. She and I leave the bay together to go look at the upland part of Alameda Creek. As we drive, she tells me that dams aren't the only infrastructure within the stream blocking water flow and sediment delivery; there are also concrete sills and weirs. What sediment makes it through that highly engineered gauntlet is shot out past the marshes into the bay. For several decades, the creek has been known by another, more industrial name: Alameda Creek Flood Control Channel.

Sediment piling up behind dams fills reservoirs, making it a maintenance problem, so managers often clean it out. This supply is a possible resource to deliver to the marshes, permits willing. But scooping it out and trucking it to the marshes is slow and expensive. Opening up the dams and allowing the sediment to flow through would be faster and easier. In some cases, existing dams could be operated to mimic natural, periodic pulses that flush sediment out to the coast. Or if some dams need to be replaced, newer designs could spill water from the bottom, allowing sediment to pass.

We arrive at the dead end of a quiet suburban street in Union City, where a levee looms above us. Parking at Beard Staging Area on the Alameda Creek Regional Trail, we walk up a steep hill to reach the path atop the levee. From that vantage point, we look down into the V-shaped concrete flood-control channel. At the bottom, riparian plants—yellow, beige, green, red—obscure a thin trickle of water.

It's not just the sediment that's blocked from moving through this creek. Humans are too; we can't reach the creek from here. There's also no easy way for people from this neighborhood to visit other people's homes just across the creek, perhaps a hundred yards away. People run or walk dogs along this trail, "but being able to get down into the creek is a way different experience than standing on top of the levee," Morgis says with feeling. "People used to fish in Alameda Creek. Older residents were really engaged with it."

Her team wants to change that, building paths down into the channel to reconnect people with their local ecosystem, allowing them to "meander within the creek channel itself." This goal is why her team dubbed itself Public Sediment. Reuniting people with each other, their local water, and other natural resources in their backyards "builds more empathy for ecosystems and builds a constituency, a group of people that could advocate on their behalf," Morgis explains.

People working on the Eden Landing restoration are excited about this plan to enable freer water flow from the creek to the marshes under restoration. Aside from delivering much-needed sediment, improving water flow could also make it easier for fish to pass. And fish who are able to reach a healthy marsh would be able to grow there more safely before moving out to sea. In 2019, a state senator secured $31.4 million in funding for the project.

But because the flood-control channel is federally managed, it's complicated. Bourgeois, the former manager of the salt ponds restoration, tells me, "To punch a hole in the levee, you have to literally decommission it, which is a big process." The hydrology is complicated too, adds Halsing, the current manager. They are reconnecting part of the creek that was diverted long ago, so they need to ensure that change won't cause upstream neighborhoods to flood during extreme weather.

Still, the decommissioning is in the works, in part because the flood-control channel is no longer needed to protect salt companies'

interests. Some dams along the creek are already being decommissioned because they're seismically unsound, posing a serious safety
risk during earthquakes, or because they're silted up, reducing their
capacity to hold back floodwaters. Nevertheless, Halsing estimates
it will be a decade before the channel is fully decommissioned. Ultimately, scientists and landscape designers such as Morgis expect
that a more naturalized Alameda Creek and a sediment-fortified
marsh will enhance both human flood protection and nonhuman
habitat.

Hungering for Sediment in the Mekong

Upstream dams blocking fresh water and sediment are big problems on the other side of the Pacific as well, in Vietnam. I saw this
landscape in winter 2015, while reporting on water issues along the
Mekong River. Driving down a highway from Ho Chi Minh City,
vast expanses of lime-green rice undulate perhaps twenty feet below
the road. The fields are occasionally studded with tiny buildings—
mausoleums for deceased farmers who want to spend eternity in the
fields where they spent their days, according to a local guide.

This is the storied Mekong Delta, land sliced into wedges by
braids of its eponymous river. Home to more than twenty million
people, the majority of whom farm and fish for a living, it lies barely
above sea level, about twelve feet elevation at its highest points. It
has long been dominated by the rhythm of water. In the wet season,
upstream floods bring replenishing silt and fresh water, sometimes
covering half the delta; during the dry season, seawater from the
South China Sea and the Gulf of Thailand pushes twenty-five to
forty-one miles upstream, rendering more than 3.4 million acres of
land too salty to farm. In years with low rainfall and river levels, salt
water intrudes another twelve to nineteen miles inland.

Like the San Francisco Bay Area, the Mekong Delta provinces are
threatened from both ends. The latest projections suggest that much
of the delta could be inundated by sea-level rise and subsidence as
early as 2050. At the same time, large hydropower dams upstream,
especially in China and Laos, block waterborne sediment that the

delta needs to continually rebuild its land. The situation is already dire, yet more dams are planned: eleven on the main stem of the river and 120 on its tributaries. In 2014 scientists estimated that existing dams and those under construction would trap just over half of the Mekong's silt. If all of these additional dams are built, they projected, 96 percent of sediment will be blocked before reaching the delta.

Upstream neighbors are also beginning to irrigate more, removing water from rivers. Withdrawals, dams, and drought all decrease the surges of fresh water that push the salt back out to sea, keeping agricultural areas open for farming. What we do on land really matters: dams and sand mining have a larger impact on salt intrusion than sea level rise. Because Vietnam is at the end of the river, it must reap what its upstream neighbors sow and cannot rely on them to deliver silt or water on its schedule; those countries tend to operate dams for their own needs. In Southeast Asia, the Mekong River Commission is an intergovernmental organization that was formed to help the six countries that share the Mekong make collaborative decisions, but it hasn't been terribly effective. A glimmer of progress came in October 2020, when China agreed to share data on year-round water levels and dam releases. The Mekong Dam Monitor, a project of the Stimson Center think tank, and the remote sensing firm Eyes on Earth will keep watch on the situation via satellite.

For decades, Vietnam has also been harming the delta's water cycles in an effort to maximize rice production. Half the rice sold in country is grown here. In some parts of the delta farmers are churning out three crops a year, including during the dry season. That has required intense hydrological engineering: around 57,000 miles of canals and more than 12,400 miles of levees. This engineering has deprived parts of the delta of the regular floods that restore sediment and nutrients. The result is that farming now requires more fertilizer, and farmers pump groundwater to irrigate rice in the dry season, which draws down coastal aquifers, creating space for more saltwater intrusion and causing subsidence. That sinking land is increasing the impacts of sea-level rise. A 2019 study estimated overall subsidence at almost a half inch a year, with some areas dropping by two inches.

Here too are the Dutch, with their nose for water trouble. Together with Vietnamese water managers, they wrote the Mekong Delta Plan, published in 2013, which hews to a simple principle that would be wisely applied to any human endeavor. They recommend thinking strategically about which activities best fit with natural pressures in certain areas—and to change human activities to accommodate nature rather than fight it. For several decades, growing rice was the government priority, so people did everything possible to keep out salt water and move in fresh water. Now people in the delta are first considering water's nature in a given area —fresh, brackish, saline—and soil conditions before deciding on land use.

The Dutch–Vietnamese plan led to the ongoing Mekong Delta Integrated Climate Resilience and Sustainable Livelihoods Project, which focuses on these "hydrological ecological zones." Freshwater rice, which most farmers grow, is not sustainable in some places. People are now experimenting with floating rice, hydroponic vegetable gardens, lotus root, or rotation regimes with two rice crops a year, followed by an aquaculture fish crop. Adaptability is the watchword. If people aren't locked into a specific livelihood, they can be flexible as the land and water conditions change underneath them.

As coastal areas in Cà Mau, Bạc Liêu, and Kiên Giang provinces have become more saline, shrimp farming has boomed. Unfortunately, many farmers have cut down delta-protecting coastal mangrove forests to dig shrimp ponds. Vietnam has lost nearly 60 percent of its mangroves over the last seventy years, which makes coastal communities more vulnerable to climate change. That's because the interlocking network of aboveground roots in mangroves, which give the impression of a leafy marching army, limit erosion and offer protection from storms. Without them, coasts can erode many yards each year. Mangroves can also store almost three times as much carbon dixoide as rainforests. Plus, mangroves are a breeding ground and nursery for many aquatic species and terrestrial animals and can provide economic benefits to local people.

To address deforestation and erosion, some farmers are replanting mangroves and farming shrimp among them. A separate project called Mangroves and Markets (MAM) with international funding ran from 2012 through 2020. While these heavily managed systems do not have all the ecological benefits of a natural forest, the mangroves are making shrimp farming more profitable and sustainable. Mangrove ecosystems are a natural habitat for shrimp, as well as for other species humans eat, including crab, fish, cockles, and oysters. Raising shrimp in some semblance of their natural habitat supplies them with wild food and reduces their vulnerability to disease, decreasing the need for feed, chemicals, and antibiotics. The farmers are getting organic certification, which allows them to sell their shrimp at a premium—in part because it prohibits cutting mangrove trees and encourages replanting deforested areas.

The project was also involved with the creation of Vietnam's 2019 Forestry Decree, which sets up payments for ecosystem services for sustainable aquaculture that protects mangroves. The decree builds on related efforts by government ministries in Vietnam and international conservation and development interests. For Mekong Delta provinces, it means that policies started under MAM will continue.

Marsh Momentum from the Tides

Back in the San Francisco Bay Area, the sediment supply question continues to dominate restoration discussions. People are looking for sediment not just upstream behind dams but also from the tides. One large potential source lies in the bay's deepwater ports, which require routine dredging to remove sediment deposited by the tides. Brenda Goeden, sediment program manager for the San Francisco Bay Conservation and Development Commission (BCDC), told me that, despite the vast need, some of that dredged material is carried through the Golden Gate and dumped offshore. The quantity varies a lot year to year, but the average is about 20 percent.

Her regulatory agency was one of those created in the wake of the 1960s environmental movement that arose to protest filling the bay with land-based sediment and garbage. Whether that material was dumped as waste or to create land for development, it often buried wildlife habitat in the process. To halt these harms, the commission and other government agencies passed rules against the practice. At the time, that prohibition made sense, but now that sediment is so sorely needed, dumping dirt into the ocean instead seems wasteful. To adapt, BCDC and other institutions are now refining their rules to make more sediment available to help marshes outrun sea-level rise.

The value of sediment is clear to Goeden, whose email footer reads, in elegant script: "No mud, no lotus— Thích Nhất Hạnh." Already more than 40 percent of dredged material is used for projects in and around the bay, Goeden tells me—although none of that has gone to the South Bay Salt Pond Restoration Project. The commission and other agencies now aim to maximize sediment put to "beneficial use," although if it is contaminated, it still can't be used in the bay for wetland restoration. In some cases, if the sediment can be sequestered, materials with elevated levels of certain contaminants can be used for an elevation boost. Some sediment that doesn't pass muster goes to upland projects rather than tidal marsh.

Halsing, the salt pond restoration manager, explains that he gets most of the sediment used in his projects from construction sites. Like the dredged material, it has to be checked for pollutants. Although only about 20 percent is clean enough for reuse in restoration projects, what can be used is a win-win: "We get the dirt if it's clean enough. They don't need to pay for landfill disposal."

In 2016, a bond measure approved by voters funneled a half-billion dollars to flood management and adaptation to sea level rise, including a portion for marsh repair. But to put that money to use, the regulatory agencies need to continue to modernize, say restorationists. Getting permits for projects has been a multiagency labyrinth, sometimes taking as long as a decade. Now a partnership among the agencies is working to streamline the process.

Permits aside, it's also a practical challenge to physically move the sediment to the right spot. Trucking it in is slow going, and setting up a pump and pipe to slurry material dredged from the bay into place is expensive. Fortunately, just opening the ponds to the tides is working pretty quickly, Halsing says: "Everything we've breached so far has formed marsh faster than we thought it would." Ponds at

Northern Eden Landing, breached around 2014, already have vegetation forming in them. But Dusterhoff, the geomorphologist from SFEI, warns that if restorationists aren't able to free more sediment from and into the bay, natural accretion could slow. In a few places today, existing marshes are already eroding.

*

When restorationists can get sediment, it's difficult and expensive to move it into place quickly, and doing so raises ecological concerns as well. Bourgeois's home state of Louisiana is also starting to feed sediment to its vanishing marshes; a football-field-sized piece of land disappears into the sea every hour. Between 1932 and 2016, more than two thousand square miles have been subsumed, a slow-moving disaster with multiple causes. The heavily engineered flood controls that keep the Mississippi River in a narrow channel mean the sediment—source of the river's nickname "Big Muddy"—shoots out beyond the continental shelf like a waste product. The marsh itself has been weakened, sliced by canals and dined upon by an invasive rodent, the nutria. Oil and gas pumping (just like water pumping) have deflated the underground, causing the land to sink, making it all worse.

The assaults by Hurricanes Katrina and Rita were amplified because they struck against this already critically weakened ecological infrastructure. All of these factors combined mean that the amount of sediment needed in Louisiana is much greater than in San Francisco Bay, even though the Bay Area has lost a much bigger percentage of its historic wetlands.

Louisiana's recent Coastal Master Plan, aiming to replenish wetlands, calls for eight diversions to siphon sediment from upstream of human-made blockages on the Mississippi River. A project called Bayou Dupont is already dredging Mississippi River mud and moving it through a ten-mile-long pipeline to build wetlands. The project sprays dredged sediment mixed with water directly onto the sinking marsh to raise its elevation quickly—a process euphemistically called "rainbowing."

Rainbowing sediment is basically a desperation measure to save a marsh if it's drowning quickly, says Grenier, with SFEI. Restorationists in San Francisco Bay don't want to do this on a living marsh because it could harm life there in the short term. Rainbowing also doesn't build a natural topography with its holes and channels. Instead, it fills everything to the same flat plane.

*

On an outing to the North Bay's China Camp Marsh near San Rafael, Grenier and her colleague and friend Julie Beagle show me what that flat plane fails to deliver. Then Beagle was a geomorphologist at SFEI; now she's environmental planning team lead with the Army Corps of Engineers. They both have kids and their families are friends too; a couple of kids even joined them on work field trips to study rare plants for restoration. We climb a small hill and gaze out over acres of marsh between us and the bay. They point out the visible line between the ancient marsh, about thirteen thousand years old, and the "centennial marsh," created by the pulse of sediment that the hydraulic gold miners power-washed from riverbanks in the Sierra foothills. The ancient marsh has more curvaceous channels and an expanded color palette, signifying greater topographic and therefore plant diversity than the adjacent centennial marsh. It's similar to the difference in complexity between an old-growth forest and second growth.

"The configuration of these channels determines how the marsh works," Beagle explains. It's also a map of sorts, showing where and how often fresh and salt water move through, shaping this marsh. The size of the tide determines "how frequently it's getting wetted and determines animals and plant life." She points out spartina grass here, gum plant there.

Grenier chimes in. "It's very organized. Even bird territories are organized around the channels." Gum plant, for example, is much sparser in the centennial marsh, limiting the nesting success of some bird species.

The lesson is that natural processes—and time—make the best

marshes. A marsh harrier, gliding by on the hunt, seems to validate that principle. The more we can emulate natural processes, the better the habitat quality and more services we're going to get, not just for other animals but also for people. Curvy channels are also better at slowing down floodwater, providing humans with more protection.

For these reasons, when restoration ecologists in the Bay Area can find sediment to partially fill a sunken salt pond, they employ a natural finish. Typically they stop filling about thirty centimeters below the marsh plain, then open the area to tidal and possibly river flows, allowing natural systems to finish the job. Grenier tells me the Bay Area is angling to take a "kinder, gentler, more natural-process approach [that will allow] plant and animal populations to thrive."

In the future, if ecologists need to feed sediment to living marshes to keep up with sea-level rise, they will need a delicate way to do that to maintain marsh health. Scientists are hoping to use local tides to distribute it, a process called "strategic placement."

Another SFEI geomorphologist and other researchers looked at two approaches. One would inject sediment into daily flood tides, allowing them to move it onto the marsh. Called water-column seeding, it would be more in sync with how nature does it than pumping in slurry. But it limits the hours in which sediment can be moved and, thus, its delivery rate. The other method, called shallow-water placement, would dump sediment into shallow water or onto a mudflat so waves and tidal currents can deliver sediment to the marsh in their own time. But with this approach, animals living in the mud may be smothered.

Now shallow-water placement is being tested by the Army Corps of Engineers. Scientists will monitor potential harm to those mud dwellers and measure how much deployed sediment actually lands on marshes targeted for restoration. That will help them understand how much sediment is needed and how long restoration will take. Time and its limits are ever-present in this work. That's the big question on Goeden's mind: "Can the tides and currents move sediment onto the marshes in a significant way?"

Figuring out how to make such interventions work is urgent due to global inaction on climate change and the extent to which we've stymied natural sediment delivery with urban development and water infrastructure. "Given what we understand about climate change and what our regional scientists tell us, we have ten to fifteen years to get our act together in the Bay Area with marshes before we see an uptick in sea-level rise," Goeden says.

*

Yet some people still have outdated ideas about the water's edge. From the top of the garbage hill near Ravenswood, where my mom used to go to the dump, Bourgeois points to the north, showing me a nearly 1,400-acre land parcel on the shoreline of Redwood City. Despite the looming specter of sea-level rise, the landowner Cargill wanted to develop it with condos and office buildings, but "progress" was delayed as courts decided whether it was city or federal jurisdiction. Ultimately a federal judge ruled that it was federal land protected by the Clean Water Act and therefore could not be developed. In April 2021, Cargill said it would not appeal. Now conservationists aim to add this parcel to the restoration zones.

The good news is that preventing such doomed development and protecting marshes also means additional flexible space for both local critters and people. In my family, the tradition of enjoying space by the water continues. Bourgeois and Halsing's work on the salt ponds restoration is paying off for my brother, who often takes his kids for walks at Baylands Park near Mountain View. "I find it very relaxing to be there," he says, waxing poetic about the unobstructed views and the breeze through the reeds. "I've seen pelicans, herons, shorebirds, and beautiful sunsets." The kids also seem to calm down next to the water. He adds, "It's great because there are no cars, so they have a little more freedom to wander." One day, in turn, their kids may find respite here too. But only if we give the marshes the time, space, and sediment they need to outpace sea-level rise.

· 10 ·
Our Shared Future

LIVING WITH WATER

Nearly a decade has passed since Superstorm Sandy swamped New York, sounding a wake-up call for the iconic city of eight million, with its dense communities and pricey development at the water's edge. Climate change has arrived, Sandy told New Yorkers, and you are dangerously exposed.

City agencies, developers, and NGOs have spent the intervening years floating plans and tweaking shorelines, but the results to date are decidedly mixed and the threat relentless and existential. Around the Battery on Manhattan's southern tip, land created by filling wetlands, the sea laps nine inches higher than it did in 1950, raising the risk of both sunny-day flooding and destructive storm surges. While decision makers have fiddled, sea-level rise has accelerated. It took forty-eight years to add the last six inches at the Battery, but the US Army Corps of Engineers projects that the next six could mount in just fourteen years.

Some plans for saving Gotham call for maintaining or restoring natural protective areas, reversing the development trend that destroyed more than 85 percent of the city's historic wetlands. The project Living Breakwaters, by New York landscape architecture firm Scape, is building reef habitat around Staten Island's south end to protect the shore and serve as home to a billion oysters to clean the water. Jamaica Bay Wildlife Refuge has been replanted with more salt-tolerant trees and bushes to enhance its ability to insulate nearby homes. Paerdegat Basin Natural Area in the Canarsie neighborhood of Brooklyn has been restored from a dump site. Yet these projects are small and piecemeal given the scope of risk the city faces, and NGOs such as the Waterfront Alliance and the Natural Areas Conservancy are advocating for more.

At the same time, the city is building seawalls and mulling strategies, such as a six-mile-long mechanical barrier that would close off much of New York Harbor. This option seems more like climate-protection theater because it's designed for just 1.8 feet of sea-level rise through 2100—rather than the six feet possible cited in a state report. Also, after $119 billion invested and twenty-five years of construction, it would not prevent sunny-day flooding or cope with storm runoff. The perils of ignoring the latter were made evident by 2021's Hurricane Ida. Heavy rain caused flash floods that drowned multiple people in their basement apartments.

Only a few neighborhoods on Staten Island have accepted the move that's becoming increasingly common worldwide: retreat. Not all human habitat can be saved with Slow Water projects. In some cases, water is demanding more space than we have left open. After twenty-four people from Staten Island died in Sandy's floodwaters, homeowners in Oakwood Beach, Graham Beach, and Ocean Breeze took buyouts facilitated by the state's Office of Storm Recovery and moved away from the coast. Their properties are being returned to nature, preventing other families from moving into harm's way and making room for the salt marsh—temporarily vanquished by development of their homes earlier in the twentieth century—to regrow and create a buffer for the homes that remain. More than six hundred houses have been removed, native plants are growing, and geese, deer, opossums, turkeys, rabbits, and raccoons are settling in.

In these neighborhoods, the communities came together to advocate for this change. Their enthusiasm really intrigued Liz Koslov, a sociologist at the University of California, Los Angeles, who's writing a book about the Staten Island retreat. Because development here took off around the time a bridge was built to the island in the 1960s, local people remember when low-lying neighborhoods were largely marsh, when the house was on stilts and you could tube from it through creeks and marsh to the shore, and Grandma fished for crabs and eels off the deck. For decades, people had been fighting further development as creeks and wetlands were filled in because they associated it with the increasing flooding they were experiencing. "There's a respect for nature and a respect for the power of the water and the sense that these landscapes used to be much more fluid than they had become," Koslov says. The water just has nowhere left to go, one person told her.

The impetus that spurred Staten Island residents to retreat is happening elsewhere too, as people choose to leave after a disaster—or repeated disasters. But now scientists and government agencies are calling for a more deliberate, organized—and, in the long run, much cheaper pullback. It's called strategic or managed retreat because it's planned, as opposed to crisis driven. It relinquishes the idea of control, of "holding the line," in favor of accepting nature's power and giving it space to do its thing with less harm to humans.

After decades of resistance to the idea, which detractors characterized as "giving up," some communities are embracing it, as those on Staten Island did. Today, managed retreat driven by pragmatism is an increasingly accepted component of government policies. The United States has a small-scale history of managed retreat along midwestern rivers, where entire towns have relocated to higher ground. And now, the United Kingdom is planning a countrywide step back from the sea. With its thousands of miles of coastline exposed to the rough North Atlantic Ocean, it is mapping out where it will cease trying to hold back the sea within a decade or two. And in some places, new amphibious designs—harkening back to the lifestyle of Iraq's marsh dwellers—embrace a way to live with water without courting regular disaster. Both managed retreat and amphibious housing are the ultimate expression of Slow Water thinking, of accepting and working within what water wants.

Freedom in Facing Reality

By 2100, high tides will likely inundate land that's home to between 190 and 630 million people worldwide. The range depends on whether humanity slashes carbon emissions by midcentury or, instead, continues to fail. There is no longer any question that water is moving in and people must begin to move out. As A. R. Siders, an expert in managed retreat at the University of Delaware's Disaster Research Center, puts it: "Fighting the ocean is a losing battle." Commenting on a policy paper she coauthored in *Science* in 2019, she opined, "The only way to win against water is not to fight."

Siders and her coauthors make the case for moving away from coastal areas before things get dire. In most cases, repeated flooding is the trigger that pushes communities to move. But when emergency evacuations and destroyed homes are the catalyst for moving, retreat is inefficient, chaotic—and often inequitable, according to Siders. For example, census records show that land loss in coastal Louisiana and repeated hurricanes and flooding drove away half the residents of some towns between 2000 and 2010. Typically, the poorest people are left behind, often because they are less able to navigate bureaucracy to get the assistance they need to relocate.

Since retreat from dangerous areas is often inevitable, Siders and her coauthors argue, doing it proactively allows us to avoid death and minimize human suffering—and make the most of opportunities associated with such big changes. "We can do that the hard way, by fighting for every inch and losing lives and dollars in the meantime," Siders contends. "Or we can do it willingly and thoughtfully and take the opportunity to rethink the way we live on the coasts."

Planning proactively and watershed-wide, rather than town by town, can sometimes reduce the need to retreat. One of the easiest solutions to human-water conflict is to stop building in places that are likely to flood or to suffer water scarcity. That may sound obvious, but it's an opportunity that's still being squandered. One in ten new homes built in England since 2013 sit on land at the highest risk of flooding, according to a *Guardian* investigation. Even post-Sandy, New York continues to build in harm's way. In Red Hook, Brooklyn, I visited an area built on marshes that flooded to ten feet during Sandy. Now new condos are going up at the water's edge. Williamsburg, Greenpoint, Long Island City, and Hallets Point are also seeing new developments in locations that Sandy submerged.

Some governments are starting to say no to developers, as in

Chennai, India, where a court ruling now forbids future building on wetlands. In California, where water scarcity is a problem, a pair of 2001 laws required large developments to conduct a water supply assessment and mandated that local governments document sufficient supplies when mapping out large residential subdivisions. A town in Utah imposed a moratorium on new development in 2021 due to a dwindling water supply. Although such regulations can be unevenly enforced, places that hew to them are likely to see their foresight pay off financially. Researchers of a 2019 paper for *Nature Sustainability* calculated that every dollar invested in buying undeveloped land for conservation could save $2 to more than $5 in avoided flood damages.

When existing development floods, some misaligned policies thwart thoughtful retreat. Historically, US federal disaster relief required people to rebuild in the same location—even if it was in a floodplain. That's now beginning to change. The Federal Emergency Management Agency (FEMA) and the Department of Housing and Urban Development are starting to allocate money for relocations. The Army Corps of Engineers, in a pivot that breaks with its doctrine, now requires that local governments seeking buyout funds agree to play hardball with residents, using eminent domain if necessary to move people out of areas that the corps wants to stop defending against floods.

Canada is already going down a similar path, trying a tough-love approach after disasters. Rather than giving people money to rebuild in place, federal and provincial governments are experimenting with making disaster relief a one-time deal. Or to avoid making taxpayers shoulder inordinate costs for expensive, risky homes, in some cases buyouts are capped, rather than based on a property's market price. Unsurprisingly, such policies are not universally popular; but they can offer people relief from the trauma of repeated flooding. And removing flood-prone houses gives water more space during the next storm, protecting the buildings that remain further inland.

Canada can set such restrictions, while the United States cannot, Siders explains to me on the phone. In the United States, there's an expectation that the government must help protect your property or else it is taking it without just compensation. That's what the midwestern farmers successfully argued in that court case mentioned in chapter 7 when the Army Corps set back their levees. Canada does not have this legal precedent. But what's the alternative? The US government buying everyone out? The Union of Concerned Scientists estimated that the United States has $1.07 trillion worth of homes and commercial buildings at risk of chronic flooding by 2100. Buying them out is theoretically possible, but political and social will are major factors. Creative approaches might go down more easily—or not. California's state legislature proposed buying doomed properties, then renting them out until they were no longer usable, to recoup some funds. Unfortunately, Governor Newsom vetoed it.

Decisions on who and how to move are fraught, though. People who lack money or power often get shortchanged in these decisions or are overlooked entirely. For example, some Alaska Native and Native American communities are already seeing flooding, or damage to drinking water and sewage treatment facilities from rising waters. But they don't have money to move. Some towns and tribes seeking funds to relocate have sued oil and gas companies or the US government for their roles in causing climate change. But so far, financial assistance has been limited and the fight time-consuming. People on coral atoll island nations, including the Marshall Islands, Kiribati, Maldives, and Tuvalu, will also have to move. For them, it seems, there's little choice, just a question of when.

Retreat is never easy. I know this close at hand; fascism and genocide forced my father, uncle, and grandparents to flee their country and continent. For my grandparents, that meant starting over from scratch in middle age. They had to give up the small business they'd founded and leave behind their home, belongings, and money. In the United States, my grandfather took a job washing dishes and studied English at night. In addition to such hardships, there's often a deeper toll. When people are forced to leave their homes, they must say goodbye to the structures that sheltered them, perhaps for generations; a landscape they loved; and often, their community. "It's a complicated mix of psychological, economic, and social issues," Siders acknowledges. But with dramatic changes can come opportunity for new beginnings, as my father ultimately found in the United States; it's likely that I exist today because my family escaped. Siders emphasizes the upside in her work: "Retreat can't be just about avoiding risk: it needs to be about moving towards something better," she says. "Retreat is a tool that can help achieve societal goals like community revitalization, equity, and sustainability if it is used purposefully." When retreat is planned, there is a greater opportunity for communities to move forward together.

History of Retreat in the US Midwest

Some of the oldest examples of managed retreat in the United States, dating back to the 1930s, center around such community planning. They are not on the coast but along rivers in the Midwest, where the massive 2019 floods were only the latest and worst in a long history of successive sogging. In 2019, geologist Nicholas Pinter, associate director of the Center for Watershed Sciences at UC Davis, and his then-student James "Huck" Rees hit the road to study midwestern towns that had decided to move away from rivers together. Typically, repeated floods were the motivating factor, Pinter tells me. "Pattonsburg [Missouri] flooded thirty-two times. And they said, 'OK, enough's enough. We're gonna move the whole community up to high ground.'"

In reviewing nearly a dozen towns' moves, Pinter and Rees found some common threads. Successful community relocations tend to have a strong leader who forms a plan to move while "feet are still wet." And these relocations are more likely to be smaller towns of several hundred people up to a couple of thousand. "It's unimaginable that you're going to move Manhattan or Oceanside, California, wholesale," Pinter acknowledges.

It's also easier to do in places that have been depopulated in recent decades. Parts of the Midwest have lost half their residents over the last century or so due to farm consolidations. That means fewer folks to move and more space in which to relocate. Despite the world's booming population, that's not an unusual situation. In general, countries are urbanizing, and globally, people are pulling back from marginal farmland. Pinter's research found that if you had willing sellers for open land in floodplains, such as that around St. Louis, the land could be reopened to flooding and relieve pressure on concentrated urban areas: "Hydrologically, that's feasible."

Some towns made their decision to relocate by examining the bottom line. "I'll tell you one case study I kinda like," Pinter tells me conspiratorially: "Soldier Grove, Wisconsin." The town along the Kickapoo River flooded catastrophically and repeatedly during the 1970s and early '80s, he says. When the five hundred or so residents sought congressional help, the Army Corps of Engineers proposed a $7 million levee—triple what the at-risk buildings were worth—to protect the heart of the downtown. The people of Soldier Grove balked. "[They] said, 'Whoawhoa*whoa*! Wait a minute!'" Pinter recalls. "'You're going to spend seven million of federal money to protect downtown, whereas the total at-risk infrastructure was assessed at something like two million dollars?!'" Additional bills for long-term upkeep of the levee "would have bankrupted the town," so they relocated instead. "It would have taken their complete tax base just to maintain the levee system. And they said no!"

In the long term, managed retreat usually pencils out over old-fashioned control. After the big Midwest flood of 1993, which killed fifty people and caused $15 billion in damages across nine states, an Army Corps study found that reducing the damage in a future similar flood would cost $6 billion in levee improvements—almost thirty times more than voluntary buyouts to remove buildings from floodplains.

Water Scarcity Too

It's not just too much water that pushes people to move; extended drought and water scarcity can also be a catalyst. For example, Syria's civil war has complex roots, but one factor is water scarcity. It was the driest of perfect storms, starting with the area's worst drought in at least five hundred years. Dams built for irrigation in recent decades by upstream neighbor Turkey also decreased flows in the Tigris and

Euphrates Rivers, reducing surface water availability in Syria, just as it did for Iraq's Mesopotamian Marshes. Plus government policies incentivized water waste in agriculture by encouraging rapid expansion of crops irrigated with unsustainable groundwater.

By 2011, failed crops, decimated livestock, and rising costs for fuel to pump groundwater pushed hundreds of thousands to 1.5 million people off their land. They moved into cities that were already bursting with more than a million Iraqi refugees fleeing the US-led war—a situation that was already causing Syrians severe economic stress when I visited in 2007. Syria's devastating, decade-long-and-counting civil war is a sobering example of the chaos that can ensue when water scarcity aggravated by climate change and poor water management collide with weak governance.

Water scarcity has people on the move elsewhere too. For example, severe drought from 2014 to the present wiped out harvests and killed livestock in Central America, especially Honduras, Guatemala, and El Salvador, making millions go hungry and forcing people off their land. Many joined the migrant caravans pushing toward the United States since 2018.

More can be done in parched places to improve rates of water conservation and reuse. Unwise water use is making the impacts we feel more extreme. But we may ultimately have to retreat from some arid areas as they get drier. Massive desalination and water-engineering projects have made possible the population booms in desert areas like the Persian Gulf and southwestern United States. But these places may not be able to sustain those numbers of humans in the long term. Several Gulf countries have nearly tapped out their groundwater to supply their growth to date. The US Southwest remains in a megadrought going on two decades and counting, the driest period since the late 1500s. Human-caused climate change is a factor, scientists concluded in 2020.

The future may see migration away from these water-insecure places toward those that are rich with water, such as the United States around the Great Lakes, Canada, Russia, Scandinavia. Unfortunately, one early outcome is not moving but water grabbing. And this phenomenon is not exclusively limited to targeting poorer countries such as Kenya. Saudi Arabia has bought up ten thousand acres in Arizona to grow hay to feed cows at home with subsidized, unsustainable water.

In England: Marshes to Farms to Marshes

One place taking both managed retreat and regional planning seriously is the United Kingdom, with a national program called coastal realignment. In England alone, the government has identified 1.8 million properties at risk of coastal flooding and erosion by 2080, along with £120–£150 billion (US$169–$212 billion) worth of infrastructure such as roads, schools, and railways.

The UK Environment Agency is acknowledging that it cannot win a war against the sea, even if it could afford to wall off the whole coast—which it can't. Instead, it is planning strategic pullbacks. That means abandoning or removing hard-barrier coastal protections—generally, seawalls or embankments of big rocks—in places where they must be maintained constantly and still sometimes fail. Instead, the agency is suggesting that people move back from threatened coastlines. One strategy is to build new barriers against the sea further inland and allow marshes, estuaries, and other coastal habitats to migrate into the breach. Because the new barriers are buffered by natural ecosystems that dissipate wave energy and reduce erosion and flooding, they are less expensive to maintain and less likely to be undermined.

Most planned projects are in areas of low human density, such as marginal farmland, rather than urban zones. But depending on the location, the scale of projects, and the physics, these areas can reduce impacts on nearby urban areas. Several such projects have already been built in the UK.

On a cold, bright day in early March 2020, after saying goodbye to the UK beaver believers in chapter 4, Peter and I drive to the sea. I want to see the Medmerry Managed Realignment Scheme, the largest planned retreat on an open coast in Europe at the time it was completed in 2013. Now it is a nature reserve administered by the Royal Society for the Protection of Birds.

Two hours southwest of London is Selsey, a town of about ten thousand people in south Sussex that lies on a peninsula called Manhood. (Yes, really, but it derives from *Man wode*—"wood"—and refers to a long-gone communal forest.) Selsey lies at the tip of the peninsula, poking its nose out into the English Channel. The west side of the nose is home to the post-realignment Medmerry Nature Reserve. Selsey is a typical English seaside town with pubs, charity shops, and grocery stores lining the quiet high street. Low-lying and historically riven by estuaries and rifes (a local word for creeks), Selsey was, in fact, an island prior to about 1750.

Residents accepted the town's tidal separations until about two hundred years ago, when the lord of the manor decided he didn't like getting a ferry across the marshes and wanted greater profits. So he drained the land, reclaimed it from the sea, and turned it into farmland. Now there's just one road in and out of town, via a bridge over an inlet called "the ferry channel," where the boat used to take people across at flood tide.

Although Selsey lies within the uneasy embrace of the sea, humans have lived here as far back as the Stone Age. The town was once the capital of the Kingdom of the South Saxons, and its name, meaning "seal's eye," or "isle," is derived from the Saxon language. Pete Hughes, an ecologist with the nearby Chichester Harbour Conservancy and a consultant on the Medmerry project, tells me that "a huge amount of archaeology was discovered" during the managed realignment project.

The range of findings speaks to how the geography here has changed over the centuries, shifting between land and water as the seas gradually fell and rose as ice ages descended and receded—and now are rising rapidly with climate change. They illustrate how many of the landscapes we think of as fixed are ever in motion. Archaeologists found a submerged oak forest from Neolithic times (2455–2290 BCE), now visible at low tide. They also discovered an Iron Age (760–410 BCE) human skull held by a rock, its spinal column trailing out in the water behind. Per human practices of the time, the body was likely placed on a wooden platform in a creek, once in this spot. More recent artifacts from about five hundred years ago show how fundamental the sea has been to human life: fish traps, an eel basket, shrimp and prawn traps. Archaeologists also found two water wells. One from the mid-eighteenth century looks like a wine barrel with its top missing, the inside lined with bricks. The other is lined with breeze blocks of chalk. When the coast was further out, "these wells would have supplied fresh water," says Hughes.

In fact, the name Medmerry comes from a small village that disappeared into the sea a few hundred years ago. "It was built on marshes," explains David Rusbridge, whose family farmed on the Manhood Peninsula for four hundred years. Medmerry means "middle eye (isle)" because it too was once an island, as revealed on a 1752 map. A neighboring hamlet, East Thorney, was also lost to the sea, possibly done in by nineteenth-century drainage ditches to move water out of the marshes. Without those marshes as protection, the sea moved in around 1900. Now the coast is where the marshes once lay.

In modern times, before the Medmerry Managed Realignment Scheme, this stretch of shore was the most threatened by coastal flooding in southeast England. It was rated a 1 in 1 risk, meaning it was likely to flood every year. During a severe flood, if Selsey's one access road were inundated, people could be trapped there, with no electricity, sewage treatment or emergency services. Other nearby communities were also at risk: a settlement of small prefab, portable vacation homes to the west of Selsey called West Sands caravan park, and Bracklesham Bay, a village to the northwest. These communities were protected by a "shingle embankment"—essentially, a low barrier made of rocks—erected where East Thorney once lay.

Every winter, workers would go out to the shore with bulldozers and push up the rocks afresh, a job that cost £300,000 (about US$413,000) annually. In 2008, these efforts failed, and thirty people had to evacuate when the sea came in. Damages cost £5 million (US$7 million). In 2010, another tragedy struck when a sixty-year-old bulldozer operator was crushed by his machine when he stepped out for a bathroom break. Pippa Lewis, an environmental project manager with the UK Environment Agency, tells me, "Work during winter storms to maintain the embankment was dangerous, expensive—and wasn't really working, to be honest."

The goals of the Medmerry realignment were to make the local communities safer and to recreate lost coastal habitat that would benefit endangered species. The Environment Agency bought land from farmers just behind the shingle bank so the marshes could reclaim space that was once theirs. "From an historic perspective, a lot of what we are realigning is just land that we have claimed from the sea," Hughes explains. "There's a sense of . . . we stole it from salt marsh two hundred years ago; we should be putting it back."

To see some of the 740 acres of habitat, Peter and I decide to rent bikes. At a small Selsey shop called Peddle Wise, we opt for cruisers over mountain bikes because they have fenders. A drenching rain the day and night before has left ponded water scattered across the flat terrain; the fenders will prevent large waves of water from churning up onto us.

We set out past an old windmill, then cruise through West Sands caravan park, which seems mostly empty at this time of year. At the edge of West Sands, we come to the southern end of the realignment project. Here, workers removed the shingle bank barrier to the sea and built a new 4.3-mile-long, U-shaped levee that juts inland 1.2 miles. The new levee is made of local clay, already held strong with grass and other plants, and fortified on the sea edge by large rocks. Standing atop the levee, we can look down into what was farmland in 2012 and is now a young marsh that floods with the tides.

The length of coastline that allows the sea to come in, more than 360 feet, is vast enough that the homes and buildings on the other side appear tiny on the horizon. Now prone to sudden inundation, the intertidal area is not accessible to the public. Instead, a winding inland path tracks the new slither of levee. The hard-packed dirt of the path is pocked with puddles this morning, and some segments are fully flooded. Reflecting the sky, their depth is impossible to gauge, so traversing them is a leap of faith. But to hesitate means stepping into water. I plow through at a steady pace, water shooting up in rooster tails from my tires, feet carefully balanced on pedals skimming just above the surface.

As we peddle, we pass through a variety of lowland habitats. In the intertidal zone, the sandbars and rocks scattered across mudflats offer chunks of land for birds who make their living here, including dunlin, curlew, gray and ringed plovers, oystercatcher, redshank. They hunt thirteen species of marine mollusks who have moved in. Just a short distance inland, saline plants are already established. Still further from the sea, reed beds give way to grassland and low shrubs.

*

David Rusbridge, the man whose family farmed in the area for four hundred years, was asked by the UK Environment Agency to sell his 350 acres so it could be returned to marsh. His cousins, who owned 400 neighboring acres, were also asked to sell. "Me and my cousins, we sat down and talked about it," he tells me. "We were in no position to argue" with the government. "They would have got their way one way or another. So our view was work with them, not work against them."

He believes it was the right decision. Rusbridge had been on the local flood defense committee, so he knows the challenges—and costs—of keeping out the sea. When the national Environment Agency decides that it will no longer protect a stretch of coast, such as the annual bulldozing once required at Selsey, then it's down to local districts to decide whether to pay to protect it. The prospect of shouldering that financial burden locally changes the calculus, in his experience.

When the Medmerry Nature Reserve project came along, Rusbridge decided he was ready to quit farming. "If I'm entirely honest, conventional farming was quite difficult to make much money on it," he says. Consolidation of family farms like his by industrial companies created stiff competition. And his children weren't interested in farming. The government paid him market price for his land, and given the tough economics of farming, "it seemed a reasonable way out."

Still, after four hundred years of family history here, "it marked the end of an era," he adds a bit wistfully. "But we were happy that it was going to go to a good environmental cause." Both of his adult children work on environment issues. Since selling his land, Rusbridge has started a self-storage business in Bracklesham Bay. "Self-storage pays a hell of lot better than farming," he laughs. He still has friends who farm, and "when they're struggling in the harvest, when it's raining every day, I'm glad I'm not doing it."

Part of Rusbridge's former farm is now a section of the 452 acres of intertidal area Peter and I gazed across from the path atop the levee. Creating this habitat was a key goal of the project, and mitigation funds to compensate for similar habitat lost elsewhere to development helped pay for it. Salt marshes are not just great buffers against wave energy, they are also vital nurseries for fish and serve as home and pantry for other wildlife.

In addition to the intertidal areas (mudflat, salt marsh, and transitional grassland), Medmerry has freshwater and terrestrial habitats. They transition from one to another in the way that San Francisco Bay marsh restorationists are championing. The habitats are helping the UK to meet various national and international targets for biodiversity protection. Already the Royal Society for the Protection of Birds has seen breeding populations of avocet and little ringed plovers. Ducks such as teal and widgeon are prevalent. Creatures have even more room to find what they need thanks to physical connections between Medmerry and the older Pagham Harbour Local Nature Reserve—a tidal estuary on the east side of the Manhood Peninsula that was formed about a hundred years ago when another barrier wall gave in to the sea.

On our bike ride, we pass small groups of people walking, sometimes with dogs, and a few other bikers. At one point, near a paddock with grazing animals, birders are making a count. They tell us that the birds they see on the farmland include corn bunting, reed bunting, linnet, skylark, gray partridge, and yellowhammer, the poster bird of Medmerry Nature Reserve.

While it's hard to give up one's land, the Medmerry project seems to have provided a happy ending for Rusbridge and his family. But it's not clear whether other landowners in the path of coastal realignment will do as well. Medmerry was a flagship site, and the Environment Agency invested £28 million (US$38.5 million). But the agency likely can't afford the same for every coastal spot at risk. While it will continue to look for opportunities to "manage realignment" in places that can create habitat for wildlife while reducing risks for people, some sites may be allowed to breach naturally, relinquishing control to nature without investing millions to micromanage it.

Vole Patrol

Although other local people around Selsey were less affected than the farmers who sold their land, public sentiment can be a big challenge on a project like this. "People are resistant to change and understandably anxious about something that is going to alter their neighborhood," Hughes, the ecologist, tells me. The human instinct to build a fortress—such as the shingle barrier—is strong. People tend to assume this will provide the best protection, so proposing its removal raises community concern and conflict, he warns.

Explaining the project to the community and engaging them in decisions can help. "A lot of the local community had been quite contentious at the early phases," Hughes says, "but as they realized the benefits and protection they'd be afforded, people gradually became much more positive about the project." Aside from better flood protection, the marsh has brought economic benefit. The return of wildlife and more than six miles of new paths for walking, biking, or riding horses has brought visitors to the area beyond the summer tourism season, more than quadrupling their number. Local people use the path a lot too, and the intertidal habitat creates fish-spawning and nursery areas that are helping to sustain local fishers.

After biking around to the far side of the new levee, Peter and I see a couple of freshwater lakes that attract different birds, such as lapwing and golden plover. These ponds were created both for habitat diversity and also to temporarily hold flooding from two local creeks during heavy rains. When the tide is up, freshwater can't reach the sea, Hughes explains, a phenomenon called "tide locked." Within the levee here, one-way sluices let stormwater escape to the sea when the tide is out. Another channel diverts water to the area from a creek upstream of Medmerry to create a longer-lasting fresh-water feature.

One freshwater denizen is a beloved creature in Britain: water voles were made famous by Scottish author Kenneth Grahame in his novel *The Wind in the Willows* (1908). One of the main characters, Ratty, is not actually a rat but a water vole. I had a charming phone chat with Rowenna Baker, a freelance conservation biologist who did her postdoctoral research on how a remnant population of water voles colonized new habitat at Medmerry. Although water voles are about the size of a rat, they look and swim more like miniature beavers.

Once common here, water voles are now nationally endangered due to lost habitat. "We've got rid of all our flipping natural wetlands!" Baker exclaims. The remnants are often isolated or unsuitable for voles. Another threat is the invasive North American mink, which is well established in the UK after releases and escapes from fur farms. They can swim and fit into water vole burrows. But there's hope for the voles; as Baker puts it: "They are pretty fast breeders." If people can restore a large enough area of suitable habitat free from mink, vole populations could recover.

Like beavers, these semiaquatic rodents cut vegetation that is far larger than they are: big bulrushes and reeds that they fell like trees. Because the plants are so much taller than the animals, water voles are often invisible to the human eye—unless you know what to look for. "Often you can see one piece of vegetation just drop, and it falls down onto its side, and then"—Baker whispers—"you listen really quietly. You can hear them chomping away at the vegetation."

For her postdoctoral research, Baker monitored the voles' escape from the salt water when the barrier was breached into a freshwater channel built by the project. Creating this new habitat for them worked. Genetic studies showed that many of the voles who colonized the new freshwater channel within two years of the breach were originally from the sea-flooded zone.

Baker tells me her favorite thing about water voles is that the females, who patrol their territory to keep out other females, are fighters. "When you trap a female and then pop her in a Pringles tube—because they just fit quite nicely in a Pringles tube—her tail just starts going slowly from side to side, and then it gets faster and faster and faster, and then she'll try and turn around in the Pringles tube and bite your nose off!" Baker says, with great humor. "Honestly, they're hilarious. They are so feisty."

One bit her finger down to the bone, an example of how fierce they can be. "But I love them for it!" she insists. "I'm like, you know what? Good on you! You've got the character of survivors, d'yaknow what I mean? They're not just this meek and cute little creature. They're actually feisty little devils. And I love that in a small mammal." She laughs.

*

As the water voles make themselves at home in—and defend—their new freshwater habitat, the salt marsh is doing its job protecting human habitat. The winter after the Medmerry project was completed, 2014, was a stormy one on England's south coast. It was "some of the worst weather for twenty years, with a sustained period of very high tides, strong winds and stormy seas," according to a UK Environment Agency report. "The new flood defenses have held firm and are working as planned." Flood risk to the area, once rated one in one year, has been revised to one in a hundred. The main road to Selsey, a sewage treatment plant, and 348 homes now benefit from that increased level of protection. Now several years later, the salt marsh continues to buffer the community, according to Lewis with the Environment Agency, despite "some big, big storms."

Retreat from Medmerry may seem less than ambitious because only marginal farmland was taken out of operation. But making more space for water in places easier to give back to nature absorbs some of water's energy, reducing pressure on more-populated areas, such as nearby Portsmouth and Southampton.

Rusbridge, the former farmer, has taken this principle to heart: "I thought, perhaps that's the writing on the wall, and I ought to move into the hills." Sea-level rise predictions vary, but "if we're going to get a meter or meter and a half of sea-level rise, it's pretty much back to the drawing board, because a lot of Ham [another name for the lowland area] will be taken by the sea whether we want it to or not." He still goes down to the Manhood Peninsula often. "I walk along the trail bank, the bund, and try to recognize which fields were which," he says. "It becomes harder as time goes on." Day after month after year, water reclaims its space, obscuring human patterns.

Amphibious Adaptations

The longer we think about following water's lead and letting go of control, the more we may see creative ways of fitting land and people and water together. Iraq's marsh dwellers could spark ideas. One such brainstormer is architect and engineer Elizabeth English, who drew inspiration from her desire to help people suffering in Louisiana after Hurricane Katrina. Although the people on the Manhood Peninsula seem to have accepted moving back from the sea, it's not easy to relinquish one's home, especially when it's closely tied to community.

I heard English speak about her work at a water conference in Victoria, BC, a few years ago and called her recently to learn more. English was living in Baton Rouge, Louisiana, when Katrina slammed into the coast in 2005. She'd been working for the Louisiana State University Hurricane Center, studying wind loads on buildings. In the aftermath of the storm, she surveyed the devastation with dismay, feeling people's trauma: their upheaval of being made suddenly homeless by floods, the anguish of losing all their possessions, the irretrievable loss of communities, as displaced people scattered across the continent. Shortly afterward, on a plane, she sat next to a woman who'd evacuated her family to Omaha, Nebraska. "Before Katrina, her grandmother had this house in the Lower Ninth Ward, where the whole family, two hundred people, would gather once a year for a family celebration," English recalls. "And then she was talking about how everybody was dispersed. She started crying and she said, 'That will never happen again.'"

There must be a better way, English thought to herself.

The Lower Ninth Ward, a neighborhood that lies four feet below sea level in some places, was devastated by the flooding. It left an emotional mark on me too when I visited New Orleans in 2006. It was eleven months after Katrina, and the Lower Ninth was a ghost town. Some houses were completely gone, just a short concrete step remaining. Others you could walk into, but they were stripped to their studs, wires dangling, ruined things piled about, black mold growing. Broadly speaking, people from the Lower Ninth and other poorer neighborhoods were having trouble pushing through bureaucracy to get funds to rebuild. Richer, whiter communities were navigating the hurdles and already fixing their homes.

I returned to New Orleans eight years later to find a Lower Ninth Ward that had partly come back—aided by celebrity support from the likes of Brad Pitt. Most houses were lifted on stilts, squatting high above the street like menacing crabs. But it felt like a mixed bag of hope and foreboding: Many of the lifted homes were still lower than Katrina's eighteen-foot high-water mark.

English, now an associate professor at the University of Waterloo, Ontario, still spends her downtime in Louisiana. When she visits the Lower Ninth and other neighborhoods hammered by Katrina, she sees additional problems. Raised houses are more susceptible to wind damage. And people used to sit on their porches and talk with passersby. Now the porches float high above the street, making once-easy community life much more awkward.

In the days and weeks after she met the woman on the plane, English found herself pondering how to prevent further cycles of flooding, disconnection, and loss. What if the houses could get out of the way of the flood? she thought. Then people could move back in right away, as soon as the waters receded.

English began to research how people live with water around the world. She found that traditional communities in flood-prone areas have been floating their houses for centuries. In Thailand, people put bundles of bamboo under their homes to act as rafts when the water comes, tying buildings to trees so they don't float away. Along the Maas River in the Netherlands, some Dutch houses sit on hollow, buoyant foundations and are tethered to pillars so they can rise and fall with the rhythms of the water. This option excited English because it works for houses with electricity and running water. But the foundations are expensive. She wanted to create a cheaper solution that could work for people with lesser means.

In 2006, English founded the Buoyant Foundation Project to study the problem and posed the challenge to her students at Louisiana State University. One of them surprised her, telling her about the nearby community of Old River Landing, a fishing community on a lake alongside the Mississippi River. Because it is outside the levee system, people there had always experienced a little bit of flooding, but it was getting worse as the decades passed. In response, people started raising their houses on higher and higher stilts. One guy, an engineer and inventor, came up with the idea of putting foam blocks under his mobile home so it would rise with the water. As flooding continued to escalate, his neighbors followed his example, and now about forty houses in the community are amphibious.

She and her students at the Buoyant Foundation Project took that basic idea and tweaked it to make it workable in various cultures and ecosystems. For places like the United States and Canada, they suggest using plastic water barrels, foam buoyancy blocks, or manufactured dock floats. They put the floaters inside a structural subframe under the house. The subframe and home are affixed to vertical guideposts sunk into the ground, allowing the house to rise and fall without floating away from its foundation. The homes may be designed with coiled "umbilical lines" for water supply and electricity, and breakaway connections for gas and sewer lines. The cost of a buoyancy system is around $15–$40 per square foot.

There are no mechanical parts. The water does the work, lifting the house. "It works with nature; it's not challenging nature," English has said. "It's not saying we have to stop the flood or move the water. Water is our friend because that's what lifts us to safety." In places that still rely on annual flooding to support crops with nutrient-rich silt deposits, "that's the way people feel about water," she says: "that it's a friend." That's the same sentiment that Chinese landscape architect Yu Kongjian expressed to me regarding his agrarian childhood.

People stop seeing water as a friend usually as a consequence of trying to control nature for development or profit. In so doing, English explains, "we turn water from its traditional role of being part of the cycle of life, something that is welcomed, into something that is feared and sometimes hated. It doesn't have to be that way." Our development choices create these negative consequences and, therefore, "the bad relationship with water that many, many people now have."

To the dominant development culture, English's views are subversive. "My belief system is to let water go where it wants to go," she says. "And we accommodate it by getting out of the way. Let nature be nature and don't try to control everything." Coming at the problem from that perspective allowed her to think outside the box. She found out just how far outside when she first started talking about her inexpensive, yet comprehensive amphibious retrofits—and people laughed at her openly. She thinks part of that reaction is rooted in a cultural bias that doesn't think of women as creators. Sometimes she tests people, asking them to name three women inventors. Typically they can't. "That makes my point. That was part of the incredulity," she says. "If it were such a good idea, why hadn't a man thought of it already?"

People's resistance to her idea took a toll. "It was debilitating to be ridiculed. In public. Over and over and over again," she says. "I thought about stopping. I didn't doubt what I was doing. I doubted whether I was strong enough to keep going." But the distress she felt on behalf of people displaced by Hurricane Katrina stayed with her. So she kept at it.

English's amphibious retrofits allow people in flood-prone areas to protect their homes from repeated flooding by simply lifting them out of danger for the duration of the flood. Yet, in the United States, she's had trouble in getting permits for them because FEMA, which administers the National Flood Insurance Program, won't insure an amphibious building. And banks won't grant mortgages for uninsured buildings in floodplains.

At the same time, though, buildings behind a levee certified by the Army Corps of Engineers are not considered to be in a floodplain, so people living there don't have to follow regulations for flood protection. It's institutionalizing the levee effect, putting an official stamp on the idea that an area is safe. It encourages people to stay in a risky place by raising the value of their property with the assumption of continued government protection. "The problem is that people who made those decisions forgot there are only two types of levees," she says, quoting that old joke—which is actually not so funny.

"It's crazy" to put a wall around a floodplain and then assume it won't flood, she fumes. Flooding will be less frequent, but when the levee does succumb, the consequences will be all the more devastating because people moved into the floodplain and became complacent. She laughs in frustration. "You know, it infuriated me then and it still infuriates me. I mean, honestly, how can you have such hubris to think that, just because they put a levee there, it's not a floodplain anymore!"

English couldn't accept that the government would block greater resilience for existing homes behind levees by denying them insurance if they retrofitted. "It took me seven years," she says, "but I finally convinced FEMA that their own regulations allowed this, that amphibious retrofits could be grandfathered in" for houses built before a jurisdiction adopted the National Flood Insurance Program.

Now Canada is considering amending its national building code to include amphibious retrofits, and English is an adviser. It would be the first country in the world to do so, she says, although the provinces have their own building codes, so they would each have to approve a national code before it could be used.

Amphibiating a house (a verb English invented) is cheaper than putting a house on stilts. But English's current design can't deal with every type of building or location. The retrofit is easiest for houses that are pier-and-beam construction atop a crawl space; then she can slip the new subframe and buoyancy blocks underneath it. The design doesn't work as well or as cheaply with buildings built directly on a concrete slab, or that have a basement. It's also not robust enough to deal with strong waves, ice floes, or fast-flowing rivers. Yet. For now, English recommends it for river floodplains and the banks of lakes, and she's continuing to think about how to address these other situations. She's also taken her designs elsewhere around the world, including Vietnam and Bangladesh.

One of her main goals is to find a solution that works for people of lesser means. In places where the homes don't have electricity or running water, her retrofits would typically be less than $5 per square foot. They must be affordable for her clients and must make use of what people already have. This avoids waste—one of the downfalls of our current society, as English sees it. "Reusing an existing structure is one of the most sustainable things that can be done," as opposed to tearing down an existing building, sending a huge pile of debris to the landfill, and manufacturing all new materials, with the commensurate environmental, water, and carbon footprints. Or rebuilding a house in a floodplain several times.

She's also keen to avoid what she calls a "cultural colonization" of trying to impose upon a place an incompatible way of life designed for elsewhere. She tries to match her designs to the needs of local people. The human urge to control nature extends to "people controlling other people," she laments. Usurping the existing sustainable practices of a place, its autonomy, culture, and way of life, is akin to "the same way that nature is disrespected." And as we've seen in Iraq, India, Peru, people have developed locally appropriate, innovative ways to live with water.

Another major influence on English's thinking came during her doctorate in architectural theory, for which she studied the influence of Russian Slavophile mystical philosophy. It's a complex ideology, but in short, it underscores human ways of knowing beyond rational thought and the connections among everything in the universe. It has parallels to Indigenous worldviews, ecology, and systems theory and is considered a precursor of environmentalism. "It gave me a lot more freedom to see water in a different way and to see the human relationship with water as a part of nature, as being more symbiotic," she tells me. With this perspective, she realized that "to live in harmony, you have to have respect. And if water is being a problem, well, maybe it's because you're not respecting water enough." That puts the onus on the dominant culture to turn things around, to appreciate water rather than fear it.

There's now an annual conference on amphibious architecture. With climate change causing a massive escalation in flooding disasters, "nobody laughs at me anymore," she says with rueful vindication. Adapting to climate change and reducing flood impacts are important to her, but the heart of English's motivation still goes back to that woman she met on the plane: "I'm doing this to reduce unnecessary trauma."

Restoring Our Relationship with Water

Trauma triggered by flooding and water scarcity seems to be all around us today. The landscapes where we live are imprinted upon our psyches, freighted with personal and cultural meaning, so when droughts and floods force us to move, we suffer loss. Loss of property and possessions; loss of home and community; loss of beloved landscapes and their distinctive species, weather, sounds, and scents; loss of quality of life. That melancholy I felt as a child, watching stone fruit orchards fall to silicon chips, is a feeling that stalks many of us today.

People have clearcut trees, fished out and polluted waters, and driven wondrous creatures extinct throughout the centuries. But the rate and scale of destruction we are now causing is accelerating: glaciers melting, rivers drying up, Amazon and boreal forests falling, wild animal populations crashing.

The toll this devastation takes on our mental and emotional health is difficult to fathom, especially because the dominant culture glosses over it, a form of gaslighting. But it's real, and it's serious. Especially when it's attached to life's most vital ingredient—water. Insecurity around the behavior and availability of water is destabilizing. It is natural to feel anxiety, depression, despair, anger, grief in the face of these losses, a bleak-looking future, and the prevailing human systems that seem unwilling to change, even though the need for transformation is obvious. Climate scientists warn that we have about a decade to dramatically reduce our greenhouse gas emissions to avoid catastrophic changes like the water extremes we have already begun to suffer. Biologists implore us to protect other life on

Earth on a similar timeline if we hope to avoid cascading ecological collapse.

Yet global society presses on with "progress." Ignoring systems theory, devaluing ecosystem services, and seeing ourselves as separate from the natural world and each other all breed fear and greed—a scarcity mindset. That leads us to make short-sighted development decisions in a futile attempt to control the environment and keep the good times rolling (for some). But privileging ourselves isn't working; instead, this approach is impoverishing humanity, as extreme environmental damage reduces quality of life. Just tweaking business as usual is not going to halt this downhill train. Business as usual is what got us here.

Proponents of a movement called Deep Adaptation argue that climate chaos, followed by societal breakdown, is imminent. They are sounding that alarm not to incite panic but to inspire a conversation about what comes next. It's an opportunity to rethink everything, to prepare for radical change in ways that reduce conflict and trauma. One facet of such radical acceptance is to admit that water always wins. That admission is not weakness. Instead, it's the foundation for strength because it opens us up to innovative solutions. The way we are relating to water and the natural world is not innate. We create our narrative, and we can change it. Letting go of what is frees us to embrace what can be. It's what A. R. Siders, the adaptation expert, invites us to do: plan ahead to pivot, as necessity demands, with less sturm und drang. By shifting our relationship with water now, we can move toward something better with less upheaval.

The water detectives I met are ahead of this curve. Most are pragmatists, working to do what they can within the constraints of the dominant mindset. But they are pushing the envelope. They see firsthand that what we are doing isn't working, so they approach water with humility, curiosity, and an openness to change.

In championing Slow Water, the detectives are advocating for a fundamental shift in how we think about ourselves, our systems, and our world. They endeavor to understand water and accept it for what it is instead of trying to shape it into what we might wish it to be. In so doing, they build a partnership with water based on respect, meeting as equals. By closely observing other species' and cultures' relationships with water and valuing that knowledge, they find elegant solutions that can begin to heal many of the problems we're causing ourselves.

Imagine how much faster the water detectives could work if we, as a global society, followed their lead and asked ourselves: What does water want? If we all tried to understand water's behavior and sought to change our relationship with it? What could be if we took a page from cultures that consider water to be a relative, a friend, a collaborator, rather than an enemy or a commodity?

Kelsey Leonard, the Shinnecock citizen and professor, articulates the power of valuing such knowledge. She says that recognizing water's rights, acknowledging it as a who, with agency, "reverses the accepted hierarchy of humanity's domination over nature. As human beings on this planet, we are not superior to other beings. We are not superior to the water itself." She enjoins us to become good stewards again, aided by laws that enforce legal personhood for water, such as the rivers in New Zealand, India, and Canada that now have the right to exist and thrive. That cultural cognitive shift has logical follow-ons: honoring original treaties between Indigenous and non-Indigenous peoples for water protection, dismantling property systems that privatize decisions about water, and ensuring the well-being of water before our narrowly constructed human demands— because if water is not well, it cannot provide for us. Leonard issues a Kennedyesque call to action: "Ask not what water can do for you, but rather, what are you going to do for water?"

What *are* we going to do for water?

While decision makers and water detectives play a large role in shaping our relationship with water, we can all help shift the nature of that bond. We can hold space for water—not just on our land, but in our hearts and minds. The water detectives have taught me to cultivate curiosity about what water is doing around me. I now take time to question as I move through my world: What is that musical burbling sound? Where do the ghost streams travel in my neighborhood? What plants are growing nearby? Are they conveying messages about hidden hydrology? How does my local water want to interact with the wider watershed? How can I support and accommodate water within my community?

Considering what water wants may have sounded a bit mystical, even radical, when you opened this book. In fact, it's a practical and proven path to creating a better world in which people are happier and communities more adaptable. While Slow Water projects reduce the risk of floods and water scarcity and the subsequent anxiety those situations bring, they are simultaneously creating more dynamic, diverse, enticing habitats for us. These are beautiful spaces that people love. Think of the lush willows, alders, and salmon helping to keep Seattle's Thornton Creek from flooding homes; Devon's beavers calming the River Otter while attracting frogs, herons, and delighted humans; the productive family farms in Kenya's Highlands; my brother's tranquil walks along the rebounding marshes of San Francisco Bay.

These healthier places can also cultivate more peaceful spaces within us, and greater adaptability. Slowing ourselves to observe the nuances of our environment—the ebb and flow of water, the growth and decay of plants, the behaviors of other animals—is meditative and joyful. In these moments we step outside ourselves, stitching connections between us and the world.

If we can find the societal will to make space for Slow Water across watersheds, this healing can multiply. We can make room for rivers to slide over their floodplains once again. We can help water move underground and reunite with its aquifers. We can grant estuaries the sediment and space they need to grow. We can protect water towers. We can accept that riverbanks and shorelines don't belong to us; rather, they are in constant flux, shared with us in phases. In letting go, in providing space, we acknowledge the power of waterlands—to hold water, to hold carbon, to hold life, including us. In giving, we receive.

Acknowledgments

It takes a global village, seemingly, to write a book, and I'm thankful to the many people who helped me along the journey to capture the essence of Slow Water.

Over the years, reporting has been my tool to explore water's ways. The more I've learned—about what water wants, its relationships, how it moves—the more fascinated I've become. Some of the material in this book was developed from early articles for *Scientific American*, *Nature*, and *National Geographic*. Articles for *bioGraphic*, *BBC Future*, and *Scientific American* were adapted from my book reporting and writing, helping to finance the project. I am particularly thankful to Mark Fischetti at *Scientific American* and Steven Bedard at *bioGraphic* for their masterful editing and for the travel funding that helped get me to China, Iraq, and India. I'm also grateful to the National Geographic Society for making me an Explorer and funding research in Peru and Kenya. And I thank CUNY's Craig Newmark Graduate School of Journalism for the Resilience Fellowship that brought me up close and personal with Superstorm Sandy's legacy.

Planning international reporting trips is complicated, and I couldn't have done mine without the generous assistance of Jayshree Vencatesan and Krishna Mohan in India. In Peru, Gena Gammie, Catherine Mendoza, and colleagues at Forest Trends and Cecilia Gianella at CONDESAN were instrumental. For helping me organize my travels in Kenya, I thank Anthony Kariuki, Colin Apse,

Faith Cherop, and colleagues at the Nature Conservancy; agricultural extension officers Caroline Nguru and Sabina Kiarie; and Edna Wangui at Ohio University. Getting into Iraq was Kafkaesque, and I couldn't have done it without Jassim Al-Asadi, Araz Hamarash, and Azzam Alwash at Nature Iraq. Maggie Zanger and Megan Kelly also provided critical intel.

Thank you to the people who translated for me, including Marie Manrique, Carla Dongo, Agustin Nervi, Katya Perez, Vivian Tian, and Alexandra Rance.

Traveling the globe to research a book generates greenhouse gas emissions—an especially uncomfortable fact when its subject is so closely tied to climate change. To minimize emissions, I combined several trips. And I made a donation to "offset" the emissions generated by my travel. I didn't want to buy carbon offsets, because their impact is often questionable. Instead, inspired by Douglas Sheil of chapter 8 fame, I used an offset calculator to generate a dollar value and then donated that sum to a land conservation project that is protecting and restoring wetlands. I chose one near where I grew up in California, the Peninsula Open Space Trust's acquisitions in the Coyote Valley. The goal is to protect and restore Laguna Seca wetlands for habitat and water recharge and to free Coyote and Fisher Creeks to reoccupy their floodplains, protecting nearby communities in San Jose in the process.

It is important to acknowledge that the structures where I live and work, in what are now known as San Francisco and Victoria, British Columbia, stand in the traditional homelands of, respectively, the Yelamu, a local tribe of the Ramaytush Ohlone peoples; and the Songhees and Esquimalt Nations of the Coast Salish peoples.

To Jenelle Goudge: thank you for suggesting that I make a start by putting book work first every day. For guidance about the book industry, crafting narrative, general commiseration, and more writerly essentials, I offer heartfelt thanks to fellow travelers Cynthia Barnett, Gloria Dickie, Sharon Guynup, Lori Freshwater, Melissa Stewart, Ben Goldfarb, Maleea Acker, Christian Fink-Jensen, Madeline Ostrander, Bill Lascher, Osha Gray Davidson, Jude Isabella, and my *writing success!* listserv pals. I extend additional gratitude to Frances

Backhouse for encouraging me to make space for critters throughout, to Lisa Jackson for advocating that I highlight water's agency, and to Christie George for seizing on the allure of mystery and the water detectives. Thanks also to Lee Ferreira, who lent me a university textbook to feed my growing obsession about hydrogeology.

For reading all or parts of the book and providing invaluable feedback, edits, and much-appreciated encouragement, special thanks to Sharon Gies, Irene Fairley, Robin Meadows, David Fairley, Frances Backhouse, Ben Ikenson, Barbara Fraser, Edna Wangui, and Robert Luhn.

Intrepid fact-checker Amy van den Berg: I salute your nose-to-the-grindstone solidarity in the race to the finish and your fastidious efforts to protect me from errors large and small. Any errors that remain are my responsibility.

To my agent Alice Martell, who believed in this book instantly and has been a fierce advocate for it: thank you. My gratitude also goes to David George Haskell, for introducing me to Alice and for supporting my grant-seeking; to Scott Gast, who saw the potential of this book and acquired it for the University of Chicago Press; to my steady-handed editor, Karen Merikangas Darling, who shepherded me through the traces with positivity and grace; and to copy editor Johanna Rosenbohm for her careful reading, compelling style conversations, and consultations with her highly discerning nephews. Thanks also to Tamara Ghattas, Tristan Bates, Deirdre Kennedy, and the whole Chicago crew. Across the big pond, I am grateful to Neil Belton and Matilda Singer at Head of Zeus/Bloomsbury in London for bringing this Slow Water story to the UK and beyond.

There are not enough thanks in the world for all the many, many water detectives, named and unnamed, who gave generously of their time, expertise and spirit. You all contributed to my water education, and I tried to bring as many of you as possible into the text or the notes (which are illustrative, rather than comprehensive). Without you, this book would not exist.

On the home front, Mao Mao provided vital stress management services delivered via soft fur, cuddles, and purrs. And finally, THANK YOU to my incredible partner Peter Fairley, who supported this book every which way, from input on concept, process, and narrative; to reading, editing, and fact-checking; to emotional and household support; to celebratory dinners, hiking breaks, and more. You make everything in my life better.

Notes

Introduction

2   a species related to willows and other streamside trees: "Growing Poplar and Willow Trees on Farms," National Poplar and Willow Users Group, 2007, http://www.fao.org/forestry/21644-03ae5c141473930a1cf4b566f59b3255f.pdf.

2   bikers long followed this route: Mona Caron, "The Duboce Gateway Mural: Gateway to the Wiggle," FoundSF, https://www.foundsf.org/index.php?title=The_Duboce_Bikeway_Mural%3A_Gateway_to_the_Wiggle.

3   the local "Great Swamp": The Great Swamp spanned what is now the Alewife region and includes parts of Cambridge, Belmont, and Arlington, Massachusetts.

4   Worldwide, only one-third . . . to the ocean: G. Grill et al., "Mapping the World's Free-Flowing Rivers," *Nature* 569 (2019): 215–21.

5   sunk up to twenty-six feet from the loss of peat: "Subsidence in the Sacramento-San Joaquin Delta," USGS, https://www.usgs.gov/centers/ca-water-ls/science/subsidence-sacramento-san-joaquin-delta.

5   the McKinley River: Known as *Henteel no' Tl'o* in the language Lower Tanana. While the formerly eponymous mountain has officially reverted to its Athabaskan name, Denali, the *river* McKinley retains its colonial name.

6   the carbon dioxide . . . stored in forests: A. Nahlik and M. Fennessy, "Carbon Storage in US Wetlands," *Nature Communications* 7 (2016): article 13835; *Peatlands and Climate Change* (IUCN, 2017), https://www.iucn.org/resources/issues-briefs/peatlands-and-climate-change.

8   people who study water deeply: For example, journalist Cynthia Barnett, author of *Blue Revolution* and *Rain*, or water-policy expert Sandra Postel, author of *Replenish*.

9   the land ethic . . . Aldo Leopold: See Aldo Leopold, *A Sand County Almanac* (1949).

9   "a reverence for rivers": Luna Leopold, "A Reverence for Rivers," *Geology* 5 (1977): 429–30, http://waterethics.org/wp-content/uploads/2011/11/A-Reverence-for-Rivers.pdf.

9   an online talk in 2020: Kelsey Leonard, *Sacred Waters*, interview with River Collective, Science Chat 10, May 25, 2020, https://www.rivercollective.org/2020/05/27/science-chat-10-sacred-waters-with-kelsey-leonard/.

Chapter 1

13   John Cori: John Cori, interview with author, March 27, 2019.

13   a journalism fellowship focused on resilience: Craig Newmark Graduate School of Journalism at CUNY Resilience Fellowship.

14   Captain Jonathan Boulware: Captain Jonathan Boulware, interview with author, April 5, 2019.

14   at least 233 people were dead: M. Diakakis et al., "Hurricane Sandy Mortality in the Caribbean and Continental North America," *Disaster Prevention and Management* 24, no. 1 (2015): 132–48, https://doi.org/10.1108/DPM-05-2014-0082.

14   economic losses totaled more than $60 billion: B. H. Strauss et al., "Economic Damages from Hurricane Sandy Attributable to Sea Level Rise Caused by Anthropogenic Climate Change, "*Nature Communi*cations 12, no. 2720 (2021), https://doi .org/10.1038/s41467-021-22838-1.

14   The size increases the height . . . Hurricane Donna: City of New York, *PlaNYC: A Stronger, More Resilient New York* (2013), 11.

15   Sharon Guynup: Sharon Guynup, interview with author, March 26, 2019.

16   80 percent of the state's human-used water: "Agricultural Water Use Efficiency," California Department of Water Resources, https://water.ca.gov/Programs/Water-Use-And-Efficiency/Agricultural-Water-Use-Efficiency.

16   1.4 percent of its total economy: In 2018, the value of California's crops was $38 billion (California Department of Food and Agriculture, *California Agricultural Statistics Review, 2018–2019*, 10, https://www.cdfa.ca.gov/statistics/PDFs/2018-2019AgReportnass.pdf). The state's GDP in 2018 was $2.7 trillion ("Real Gross Domestic Product (GDP) of the Federal State of California from 2000 to 2020 (in Billion U.S. Dollars)," Statista, March 2021, https://www.statista.com/statistics /187834/gdp-of-the-us-federal-state-of-california-since-1997/).

16   agriculture accounts for 70 percent of human water use: *AQUASTAT—FAO's Global Information System* (United Nations Food and Agriculture Organization, n.d.), http://www.fao.org/aquastat/en/overview/methodology/water-use.

16   more than a million acres of Central Valley cropland was fallowed: Forrest Melton, "Mapping Drought Impacts on Land Fallowing in California with Satellite Data" (presentation, Pacific Northwest Drought Early Warning System Kickoff Meeting, Portland, February 3, 2016).

17   California lost . . . underground aquifers: Matt Richtel, "California Farmers Dig Deeper for Water, Sipping Their Neighbors Dry," *New York Times*, June 5, 2015, https://www.nytimes.com/2015/06/07/business/energy-environment /california-farmers-dig-deeper-for-water-sipping-their-neighbors-dry.html.

17   535,000 acres . . . to allow emptied aquifers to recover: E. Hanak et al., *Water and the Future of the San Joaquin Valley* (Public Policy Institute of California, 2019), 33, https://www.ppic.org/wp-content/uploads/water-and-the-future-of-the-san-joaquin-valley-february-2019.pdf.

17   atmospheric rivers . . . 10 to 40 percent more water: Michael Dettinger (Scripps Institute of Oceanography), email interview with author, December 2020; Xingying Huang et al., "Future Precipitation Increase from Very High Resolution Ensemble Downscaling of Extreme Atmospheric River Storms in California," *Science Advances* 6, no. 29 (July 15, 2020): article eaba1323; V. Espinoza et al., "Global Analysis of

Climate Change Projection Effects on Atmospheric Rivers," *Geophysical Research Letters* 45 (2018):4299–308, https://doi.org/10.1029/2017GL076968.

17 my mom told me that February: Sharon Gies, pers. comm., February 2017, San Jose, California.

18 Global economic losses . . . $76 billion in 2020: *Weather, Climate & Catastrophe Insight: 2020 Annual Report*, AON, 2021, http://thoughtleadership.aon.com /Documents/20210125-if-annual-cat-report.pdf.

18 people at risk from flooding . . . nearly a half billion by 2050: "Water and Climate Change," UN-Water, https://www.unwater.org/water-facts/climate-change/.

18 two-thirds of the global population . . . drought conditions: G. Naumann et al., "Global Changes in Drought Conditions under Different Levels of Warming," *Geophysical Research Letters* 45, no. 7 (2018): 3285–96.

18 New York has filled in and hardened around 85 percent of its coastal wetlands: "New York City Wetlands: Regulatory Gaps and Other Threats," City of New York, 2009, 9, http://www.nyc.gov/html/om/pdf/2009/pr050-09.pdf.

18–19 "Natural ecosystems . . . climate change": *Adapt Now: A Global Call for Leadership on Climate Resilience* (Global Commission on Adaptation, 2019), 12, https://gca .org/wp-content/uploads/2019/09/GlobalCommission_Report_FINAL.pdf.

19 the amount of precipitation . . . since 1958: D. J. Wuebbles et al., eds., *Climate Science Special Report: Fourth National Climate Assessment, Volume I* (US Global Change Research Program, 2017), https://science2017.globalchange.gov/.

19 climate change increased precipitation during Harvey by as much as 38 percent: M. D. Risser and M. F. Wehner, "Attributable Human-Induced Changes in the Likelihood and Magnitude of the Observed Extreme Precipitation during Hurricane Harvey," *Geophysical Research Letters* 44 (2017): 12457–64.

20 a storm of Maria's magnitude is now five times more likely to form: D. Keellings and J. J. Hernández Ayala, "Extreme Rainfall Associated with Hurricane Maria over Puerto Rico and Its connections to Climate Variability and Change," *Geophysical Research Letters* 46 (2019): 2964–73.

20 the western United States is suffering a climate-driven megadrought: A. Park Williams et al., "Large Contribution from Anthropogenic Warming to an Emerging North American Megadrought," *Science*, April 17, 2020, 314–18.

20 one factor in touching off Syria's civil war: David King et al., *Climate Change: A Risk Assessment* (Centre for Science and Policy, 2015) 120, https://www.csap.cam.ac.uk /media/uploads/files/1/climate-change--a-risk-assessment-v9-spreads.pdf.

20 killed about 585,000 people and displaced twelve million: "Nearly 585,000 People Have Been Killed since the Beginning of the Syrian Revolution," Syrian Observatory for Human Rights, January 4, 2020, https://www.syriahr.com/en/152189/. According to the United Nations High Commissioner for Refugees, 6.6 million are displaced internally and 5.6 million around the world. See "Syria: Events of 2018," UNHCR, https://www.hrw.org/world-report/2019/country-chapters/syria.

20 two-thirds could be gone by 2100: P. Kraaijenbrink et al., "Impact of a Global Temperature Rise of 1.5 Degrees Celsius on Asia's Glaciers," *Nature* 549 (2017): 257–60.

20 Just in my lifetime . . . that came before: Hannah Ritchie and Max Roser, "CO$_2$ and Greenhouse Gas Emissions," Our World in Data, 2020, https://ourworldindata .org/co2-and-other-greenhouse-gas-emissions. 15.43 billion tons of CO2 in 1971; 36.44 billion in 2019.

20 increasing the average temperature globally by two degrees Fahrenheit: NASA Earth Observatory, https://earthobservatory.nasa.gov/world-of-change/global-temperatures.

20 Oceans have absorbed more than 90 percent of the excess heat: *The Fifth Assessment Report of the Intergovernmental Panel on Climate Change* (IPCC, 2013) revealed that oceans had absorbed more than 93 percent of the excess heat from greenhouse gas emissions since the 1970s.

21 For each Celsius-degree . . . more vapor: This finding is demonstrated via the Clausius–Clapeyron relation dating back to the nineteenth century.

21 more water in the air . . . accelerating climate change: "Greenhouse Gases: Water Vapor (H_2O)," NOAA, https://www.ncdc.noaa.gov/monitoring-references/faq /greenhouse-gases.php#h2o.

21 seas could rise more than eight feet by 2100: Rebecca Lindsey, "Climate Change: Global Sea Level," NOAA Climate.gov, January 25, 2021, https://www.climate.gov /news-features/understanding-climate/climate-change-global-sea-level.

21 Weather catastrophes . . . displacement: Fiona Harvey, "One Climate Crisis Disaster Happening Every Week, UN Warns," *Guardian*, July 7, 2019, https://www .theguardian.com/environment/2019/jul/07/one-climate-crisis-disaster-happening-every-week-un-warns.

22 We have significantly altered 75 percent of the world's land area: E. S. Brondizio et al., eds., *Global Assessment Report on Biodiversity and Ecosystem Services of the Intergovernmental Science-Policy Platform on Biodiversity and Ecosystem Services* (IPBES Secretariat, 2019), https://doi.org/10.5281/zenodo.3831673.

22 less than 3 percent of land is considered ecologically intact: Andrew J. Plumptre et al., "Where Might We Find Ecologically Intact Communities?," *Frontiers in Forests and Global Change*, April 15, 2021.

22 the planet will hit around 10.9 billion humans by 2100: *World Population Prospects 2019: Volume 1: Comprehensive Tables* (UN Department of Economic and Social Affairs, 2019), 3, https://population.un.org/wpp/Publications/Files/WPP2019_ Volume-I_Comprehensive-Tables.pdf.

22 A cup of coffee requires . . . 1,018 gallons: "Product Gallery," Water Footprint Network, https://waterfootprint.org/en/resources/interactive-tools/product-gallery. One kilogram of beef requires 15,415 liters, and 1 kilogram of beef is about four nine-ounce steaks.

22 Earth could sustainably support about two billion people: Stephen Dovers and Colin Butler, "Population and Environment: A Global Challenge," Australian Academy of Science, last updated July 24, 2015, https://www.science.org.au/curious /earth-environment/population-environment.

22 human resource use may more than double: S. Bringezu et al., *Assessing Global Resource Use: A Systems Approach to Resource Efficiency and Pollution Reduction* (UN International Resource Panel, 2017), 6.

22 On the more than 25 percent . . . biodiversity is higher: Brondizio et al., *Global Assessment Report on Biodiversity and Ecosystem Services.*

22 the Indigenous Land Back movement: "Landback Manifesto," Landback, https:// landback.org/manifesto/.

23 we've filled or drained as much as 87 percent of the world's wetlands: Nick C. Davidson, "How Much Wetland Has the World Lost? Long-Term and Recent Trends in Global Wetland Area," *Marine and Freshwater Research* 65 (2014): 934–41.

23   the average population size of wild animals . . . globally: R. E. A. Almond, M. Grooten, and T. Petersen, eds., *Living Planet Report 2020—Bending the Curve of Biodiversity Loss* (WWF, 2020), https://livingplanet.panda.org/.

23   approximately 25 percent of known species . . . faster than what's natural: S. L. Pimm et al., "The Biodiversity of Species and Their Rates of Extinction, Distribution, and Protection," *Science*, May 30, 2014.

24   expected to set a new goal . . . by 2030: Patrick Greenfield, "Governments Achieve Target of Protecting 17% of Land Globally," *Guardian*, May 19, 2021, https://www.theguardian.com/environment/2021/may/19/governments-achieve-10-year-target-of-protecting-17-percent-land-aoe.

24   restoring the right 15 percent . . . since the Industrial Revolution began: B. B. N. Strassburg et al., "Global Priority Areas for Ecosystem Restoration," *Nature* 586 (2020): 724–29.

24   Species living in rivers and lakes . . . between 1970 and 2012: *Living Planet Report 2016—Risk and Resilience in a New Era* (WWF International, 2016), https://www.worldwildlife.org/pages/living-planet-report-2016.

25   Land-use change causes 23 percent of global greenhouse gas emissions: P. R. Shukla et al., eds., *Climate Change and Land: An IPCC Special Report on Climate Change, Desertification, Land Degradation, Sustainable Land Management, Food Security, and Greenhouse Gas Fluxes in Terrestrial Ecosystems* (Intergovernmental Panel on Climate Change, 2019).

25   escaping greenhouse gases could accelerate climate change: *Adapt Now*, 12.

25–26 Nature-based solutions . . . two degrees Celsius: Bronson W. Griscom, "Natural Climate Solutions," *Proceedings of the National Academy of Sciences* 114, no. 44 (October 2017): 11645–50.

26   these interventions currently receive just 3 percent of climate funding: Barbara Buchner et al., *Global Landscape of Climate Finance 2019* (Climate Policy Initiative, 2019), https://climatepolicyinitiative.org/publication/global-climate-finance-2019/. According to table A.2 (p. 30), agriculture, forestry, land-use, and natural resource management received US$16 billion in 2018 for adaptation and mitigation out of $534 billion spent, or 2.99% of the total.

26   climate change could cost the world . . . by 2050: "The Economics of Climate Change: No Action Not an Option," Swiss Re Institute, 2021, https://www.swissre.com/institute/research/topics-and-risk-dialogues/climate-and-natural-catastrophe-risk/expertise-publication-economics-of-climate-change.html.

26   dams . . . increase water demand: Erica Gies, "Do Dams Increase Water Use?," *Scientific American*, February 18, 2019, https://www.scientificamerican.com/article/do-dams-increase-water-use/.

27   Oak savannas . . . in the South Bay: "Indigenous Populations in the Bay Area," Bay Area Equity Atlas, https://bayareaequityatlas.org/about/indigenous-populations-in-the-bay-area.

27   early farmers felled up to 99 percent of them: E. Spotswood et al., *Re-oaking Silicon Valley: Building Vibrant Cities with Nature*, SFEI Contribution No. 825 (San Francisco Estuary Institute, 2017), http://www.sfei.org/documents/re-oaking-silicon-valley.

28   shifting baselines: Daniel Pauly, *Vanishing Fish: Shifting Baselines and the Future of Global Fisheries* (Greystone Books, 2019).

28   we live in a world that contains . . . 10 percent: J. B. MacKinnon, *The Once and*

Future World: Nature as It Was, as It Is, As It Could Be (Houghton Mifflin Harcourt, 2013), 34.

28   To keep the valuable stone fruit crops viable . . . farmers pumped groundwater enthusiastically: Seonaid McArthur and Cheryl Wessling, eds., *Water in the Santa Clara Valley: A History*, 2nd ed. (California History Center & Foundation, 2005), ch. 1 and 2.

29   a nearly failing "D" grade: American Society of Civil Engineers, *Dams*, 2021 Report Card for America's Infrastructure, https://infrastructurereportcard.org/wp-content/uploads/2020/12/Dams-2021.pdf.

29   a dam in Michigan failed and another was compromised: American Society of Civil Engineers, *Dams*, 27–28.

29   Most large dams around the world . . . a hundred years: *Ageing Water Infrastructure: An Emerging Global Risk* (Institute for Water, Environment and Health, United Nations University, 2021), http://inweh.unu.edu/wp-content/uploads/2021/01/Ageing-Water-Storage-Infrastructure-An-Emerging-Global-Risk_web-version.pdf.

31   Yu Kongjian: Yu Kongjian, interview with author, Beijing, China, April 2018.

31   "Trees, fungi, salamanders . . . if you believe Darwin": Erica Gies, "The Meaning of Lichen: How a Self-Taught Naturalist Unearthed Hidden Symbioses in the Wilds of British Columbia—and Helped to Overturn 150 Years of Accepted Scientific Wisdom," *Scientific American*, June 2017.

Chapter 2

34   In a landmark 1986 paper: G. E. Fogg, "Groundwater Flow and Sand Body Interconnectedness in a Thick, Multiple-Aquifer System," *Water Resources Research* 22 (1986): 679–94.

34–35 porous pathways . . . interconnect extensively in 3-D: G. E. Fogg and Y. Zhang, "Debates—Stochastic subsurface hydrology from theory to practice: A geologic perspective," *Water Resources Research*, 52 (2016).

35   Their thinking began to align with Fogg's: Some material in this chapter was adapted from Erica Gies, "The Radical Groundwater Storage Test," *Scientific American*, November 2017.

36   by 2100, the snowpack could shrink by four-fifths or more: DOE/Lawrence Berkeley National Laboratory, "Sierra Snowpack Could Drop Significantly by End of Century," *ScienceDaily*, December 11, 2018, www.sciencedaily.com/releases/2018/12/181211090639.htm.

36   we've already dammed nearly two-thirds of the world's mightiest rivers: G. Grill et al., "Mapping the World's Free-Flowing Rivers," *Nature* 569 (2019): 215–21.

37   people are pumping . . . Earth's largest aquifers: A. S. Richey et al., "Quantifying Renewable Groundwater Stress with GRACE," *Water Resources Research* 51 (2015): 5217–38.

37   we need to *increase* global food production by up to 70 percent: Mitchell C. Hunter et al., "Agriculture in 2050: Recalibrating Targets for Sustainable Intensification," *BioScience* 67, no. 4 (April 2017): 386–91.

38   in Southern California, water tables . . . once exchanged water: Helen Dahlke (University of California, Davis), interview with author, April 12, 2017.

38   depleted aquifers have unused capacity three times that of the state's 1,400 reservoirs: Helen Dahlke et al., *Recharge Roundtable Call to Action: Key Steps for Replen-*

ishing California Groundwater, compiled and ed. Graham Fogg and Leigh Bernacchi (Groundwater Resources Association of California & University of California Water Security and Sustainability Research Initiative, December 2018), 3; Graham Fogg, pers. comm., January 27, 2021.

38   storing water underground is a bargain at roughly one-fifth the cost of building reservoirs: D. Perrone and M. M. Rohde, "Benefits and Economic Costs of Managed Aquifer Recharge in California," *San Francisco Estuary and Watershed Science* 14, no. 2 (2016), https://doi.org/10.15447/sfews.2016v14iss2art4.

39   California's . . . and international markets: *2019 Crop Year Report* (California Department of Food and Agriculture, 2019), https://www.cdfa.ca.gov/Statistics/. Subtracting dairy, cattle, and flowers.

39   The San Joaquin River . . . ran dry for sixty miles: This lack of river water begins in the sixty miles between Friant Dam and Mendota. "Surface Water: Rivers End," San Joaquin River Restoration Project, https://www.restoresjr.net/restoration-flows /surface-water/#RiversEnd.

40   the water coursing through its banks is not even its own: It's the Sacramento's, via the Delta–Mendota Canal. Obi Kaufmann, *The State of Water: Understanding California's Most Precious Resource* (Heydey Books, 2019), 78.

40   "the least wild landscape imaginable": Sandi Matsumoto (Nature Conservancy), interview with author, February 2017.

40   One settler saw . . . "one immense sea": Robert Kelley, *Battling the Inland Sea: Floods, Public Policy, and the Sacramento Valley* (University of California Press, 1989), loc. 290 of 4581, Kindle.

40   various tribes, including Miwok and Yokut: Map of California tribes' traditional lands, California Indian Legal Services, https://www.calindian.org/wp-content /uploads/2015/09/indiantribesCA.png.

40   sheep and cattle . . . nonnative grasses: P. Laris et al., "Where Have the Native Grasses Gone? What a Long-Term, Repeat Study Can Tell Us about California's Native Prairie Landscapes," *Rural Landscapes: Society, Environment, History* 8, no. 1 (2021): 1–12.

41   "Houses were toppled . . . out of their doors.": Kelley, *Battling the Inland Sea*, loc. 340.

41   "their instinctively activist impulse . . . as they wished it to": Kelley, loc. 392–93.

41   late-summer flows averaged about three thousand cubic feet per second: Koll Buer et al., "The Middle Sacramento River: Human Impacts on Physical and Ecological Processes Along a Meandering River" (presentation, California Riparian Systems Conference, Davis, California, September 22–24, 1988).

41   "The river's channel could never contain . . . northern Sierra Nevada": Kelley, *Battling the Inland Sea*, loc. 266–71.

42   humans built partial walls across creeks: Seonaid McArthur and Cheryl Wessling, eds., *Water in the Santa Clara Valley: A History*, 2nd ed. (California History Center & Foundation, 2005).

43   Valley Water . . . ninety miles of local creeks: *Annual Groundwater Report* (Santa Clara Valley Water District, 2017), https://www.valleywater.org/sites/default /files/2018-08/2017%20Annual%20GW%20Report_Web.pdf.

43   at least 1,200 MAR projects in sixty-two countries: C. Stefan and N. Ansems, "Web-Based Global Inventory of Managed Aquifer Recharge Applications," *Sustainable Water Resources Management* 4 (2018): 153–62.

44   96 percent of that liquid fresh water is underground: "Where Is Earth's Water?,"
      United States Geological Survey, https://www.usgs.gov/special-topic/water-
      science-school/science/where-earths-water.

46–47 Scientists conducted test floods . . . without harming crops: H. E. Dahlke et al.,
      "Managed Winter Flooding of Alfalfa Recharges Groundwater with Minimal Crop
      Damage," *California Agriculture Journal* 72, no. 1 (2018).

47   crops were unharmed . . . below plant root zones: P. Bachand et al., "Implications of
      Using On-Farm Flood Flow Capture to Recharge Groundwater and Mitigate Flood
      Risks along the Kings River, CA." *Environmental Science & Technology* 48, no. 23
      (2014): 13601–9.

50   Weissmann's 1999 publication on the find: See G. S. Weissmann, S. F. Carle, and
      G. E. Fogg, "Three-Dimensional Hydrofacies Modeling Based on Soil Surveys and
      Transition Probability Geostatistics," *Water Resources Research* 35, no. 6 (1999):
      1761–70.

50   The Modesto paleo valley . . . hundreds of feet in depth: Amy LeVan Lansdale,
      "Influence of a Coarse-Grained Incised-Valley Fill on Groundwater Flow in Fluvial
      Fan Deposits, Stanislaus County, Modest, California, USA" (MS thesis, Michigan
      State University, 2005).

50   found a third paleo valley near Sacramento in 2017: Casey Meirovitz, "Non-
      stationary Hydrostratigraphic Model of Cross-Cutting Alluvial Fans," *International
      Journal of Hydrology* 1, no. 1 (2017).

50   it could accommodate almost sixty times more water than surrounding lands:
      Stephen R. Maples, Graham E. Fogg, and Reed M. Maxwell, "Modeling Managed
      Aquifer Recharge Processes in a Highly Heterogenous, Semi-Confined Aquifer
      System," *Hydrogeology Journal*, August 22, 2019.

51   the Sustainable Groundwater Management Act: See "Sustainable Groundwater
      Management Act (SGMA)," California Department of Water Resources, https://
      water.ca.gov/Programs/Groundwater-Management/SGMA-Groundwater-
      Management.

51   the law encourages recharge . . . when they need it: Esther Conrad (Water in the
      West), interview with author, February 23, 2017.

52   depleted aquifers . . . from ten Texas rivers: Qian Yang and Bridget R Scanlon, "How
      Much Water Can Be Captured from Flood Flows to Store in Depleted Aquifers for
      Mitigating Floods and Droughts? A Case Study from Texas, US," *Environmental
      Research Letters* 14 (2019).

52   there is enough unmanaged surface water statewide to resupply Central Valley
      aquifers: Tiffany N. Kocis and Helen E. Dahlke, "Availability of High-Magnitude
      Streamflow for Groundwater Banking in the Central Valley, California," *Environ-
      mental Research Letters* 12 (2017).

52   under a low-emissions trajectory . . . full use of it: Xiaogang He et al., "Climate-
      Informed Hydrologic Modeling and Policy Typology to Guide Managed Aquifer
      Recharge," *Science Advances* 7, no. 17 (April 21, 2021).

53   "With huge dams . . . the last to recognize it.": Marc Reisner, *Cadillac Desert: The
      American West and Its Disappearing Water* (Penguin, 1986), 51.

54   Josh Viers: Viers is also a watershed scientist at the University of California at
      Merced.

54   the levee setback project we are here to see in the Consumnes River Preserve: Sandi

Matsumoto and Judah Grossman (Nature Conservancy), field interviews with author, February 2017. Grossman was the project director.

54   winter's storms resupplied . . . two thousand acre-feet of water: A. M. Yoder, "Effects of Levee–Breach Restoration on Groundwater Recharge, Cosumnes River Floodplain, California" (MS thesis, Hydrologic Sciences Graduate Group, University of California, Davis, 2018), 51.

54   California's official flood-control policy . . . wildlife habitat: Jane Braxton Little, "When the Levee Breaks: Hamilton City Leads California in a New Approach to Managing Rivers," *Pacific Standard*, January 29, 2019, https://psmag.com/magazine /hamilton-city-leads-california-in-a-new-approach-to-managing-rivers; *Central Valley Flood Protection Plan: 2017 Update* (Central Valley Flood Protection Board, 2017), https://water.ca.gov/-/media/DWR-Website/Web-Pages/Programs/Flood-Management/Flood-Planning-and-Studies/Central-Valley-Flood-Protection-Plan /Files/2017-CVFPP-Update-FINAL_a_y19.pdf.

54–55 The shallow waters grow algae . . . within channelized rivers: Robin Meadows, "Raised in Rice Fields," *bioGraphic*, June 26, 2019, https://www.biographic.com /raised-in-rice-fields/. Jacob Katz is senior scientist with California Trout.

55   a law in 2016 that declared watersheds to be integral components of California's water infrastructure: "AB-2480 Source Watersheds: Financing," California Legislative Information, n.d., https://leginfo.legislature.ca.gov/faces/billCompareClient .xhtml?bill_id=201520160AB2480. Section 108.5 was added to the Water Code on September 7, 2016.

58   water lost to evaporation . . . annual public water supply: Gang Zhao and Huilin Gao, Estimating Reservoir Evaporation Losses for the United States: Fusing Remote Sensing and Modeling Approaches," *Remote Sensing of Environment* 226 (2019): 109–24, 10.1016/j.rse.2019.03.015.

60   "Decide in your head . . . never stop listening": O. E. Meinzer Award, Geological Society of America, 2011, https://higherlogicdownload.s3.amazonaws.com /GEOSOCIETY/d267ed61-55fa-417e-9424-83bbfcfd5414/UploadedImages /Content_Documents/Meinzer/2011Meinzer-Fogg.pdf.

Chapter 3

61   Abu Haider: His full name is Razaq Jabbar Sabon Al-Asadi. Abu Haider means "father of Haider," a common naming convention used after people have children.

62–63 Neolithic fisher-gatherers . . . three thousand years later: Suzanne Alwash, *Eden Again: Hope in the Marshes of Iraq* (Tablet House, 2013), 41, 48.

63   the population of marsh dwellers . . . a hundred thousand: Alwash, *Eden Again*, 9.

66   an island built over generations by cutting and piling reeds and mud: The marsh dwellers are not alone in this strategy: "Uro people have constructed islands out of the totora plant for hundreds of years, forming their own homeland in a lake that sits high in the Andes mountains, straddling Peru and Bolivia." Michele Lent Hirsch, "Visit These Floating Peruvian Islands Constructed from Plants," *Smithsonian*, August 13, 2015, https://www.smithsonianmag.com/travel/people-peru-live-manmade-islands-constructed-plants-180956218/.

66   sun-dried patties of buffalo poop: Ungulate poop has been and continues to provide fuel in many places that lack trees for wood burning, including on the Tibetan

plateau, by the Zuni people in what is now the US Southwest, and by settlers on the US western frontier in the nineteenth century.

67   Suzanne Alwash: Alwash cofounded the NGO Nature Iraq with her ex-husband, Azzam Alwash, who grew up in Nasariyah and spent time in the marshes with his father, a district irrigation engineer. See Erica Gies, "Restoring Iraq's Garden of Eden," *New York Times*, April 17, 2013.

67   "They . . . expressed an unexpectedly high respect for . . . preservation of the earth's natural environment": Alwash, *Eden Again*, 199.

67   some 175,000 people were forced to flee: United Nations Integral Water Task Force for Iraq, "Managing Change in the Marshlands: Iraq's Critical Challenge" (United Nations White Paper, 2011), http://www.zaragoza.es/contenidos/medioambiente/onu/issue06/1140-eng.pdf.

68   "A consistent pattern emerged . . . wetlands restoration project": Alwash, *Eden Again*, 110.

68   a subspecies of otter: *Lutrogale perspicillata maxwelli*.

68   the ecosystem . . . modified by human hands for millennia: Joy Zedler (University of Wisconsin, wetlands restoration biologist and consultant on the Mesopotamian Marshes restoration), interview with author, 2013.

68   "tending their watery garden . . . to maintain its channels": Alwash, *Eden Again*, 56.

68   marshes and lakes once again covered . . . 2,100 square miles: This figure includes both marshes and lakes (open water). The marshes covered about 1,300 square miles, as cited above, and the lakes about 800 square miles, according to Alwash.

69   A major flood . . . leaving Ur stranded and dry: Alwash, *Eden Again*. Delta depositions also contributed (47, 49).

69   Climate change is now exacerbating the dry swings: Nasrat Adamo, Nadhir Al-Ansari, Varoujan Sissakian, Sven Knutsson, and Jan Laue, "Climate Change: Consequences on Iraq's Environment," *Journal of Earth Sciences and Geotechnical Engineering* 8, no. 3 (2018): 43–58.

69–70 Before Turkey's Ataturk Dam . . . eight thousand cfs: Alwash, *Eden Again*. One cubic meter per second is equivalent to about 264 gallons (170).

70   Turkey has filled additional reservoirs . . . in 2020: Erica Gies, "The Real Cost of Energy," *Nature*, November 29, 2017, https://www.nature.com/articles/d41586-017-07510-3.

70   The Tigris River flowed around . . . eleven thousand cfs: Figure from Ramadan Hamza, University of Dohuk, senior expert on water strategy and policy, quoted in Samya Kullab and Rashid Yahya, "Minister: Iraq to Face Severe Shortages as River Flows Drop," AP News, July 17, 2020, https://apnews.com/article/dams-ankara-turkey-middle-east-iraq-9542368977c9ee0ae97fd2cc88933198.

70   the Persian Gulf: The Persian Gulf is also known as the Arabian Gulf.

70   Turkey has long scorned international water-sharing agreements: An informative graphic introduction to Turkey's transboundary water issues is available via the University of North Carolina–Chapel Hill. See the website *The Politics of Water: Water and Conflict in the Middle East*, https://waterandconflict.web.unc.edu/turkey-and-transboundary-water/.

71   there are now a lot of "big dams" . . . worldwide: M. Mulligan, A. van Soesbergen, and L. Sáenz, "GOODD, a Global Dataset of More Than 38,000 Georeferenced Dams," *Scientific Data* 7, no. 31 (2020).

71   Hoover stymied . . . razorback suckers, who are now on the brink of extinction: "Lake Mead: Razorback Sucker," US National Park Service, dateTK, https://www .nps.gov/lake/learn/nature/razorback-sucker.htm.

71   fish loss . . . along the Mekong River: Richard Stone et al., "Dam-Building Threatens Mekong Fisheries," *Science* 354, no. 6316 (2016): 1084–85.

71   Consider freshwater mussels . . . that in turn feed fish: Sharon Levy, "The Hidden Strengths of Freshwater Mussels," *Knowable Magazine*, June 21, 2019.

72   20 percent of the global population . . . left with less water: T. Veldkamp et al., "Water Scarcity Hotspots Travel Downstream Due to Human Interventions in the 20th and 21st Century," *Nature Communications* 8 (2017): article 15697.

72   until a diplomatic breakthrough in 2021: April Reese, "Amid a Drought Crisis, the Colorado River Delta Sprang to Life This Summer," *Audubon*, Fall 2021, https:// www.audubon.org/magazine/fall-2021/amid-drought-crisis-colorado-river- delta-sprang.

72   the controversial Inga 3 dam . . . who lack access to electricity: Grace C. Wu and Ranjit Deshmukh, "Why Wind and Solar Would Offer the DRC and South Africa Better Energy Deals than Inga 3," Conversation, July 23, 2020, https:// theconversation.com/why-wind-and-solar-would-offer-the-drc-and-south-africa- better-energy-deals-than-inga-3-142411.

72   Zambia . . . into an economic crisis: Erica Gies, "Can Wind and Solar Fuel Africa's Future?," *Nature*, November 3, 2016, https://www.nature.com/articles/539020a.

73   "emissions from tropical hydropower . . . for decades": P. Fearnside and S. Pueyo, "Greenhouse-Gas Emissions from Tropical Dams," *Nature Climate Change* 2 (2012): 384.

73   a liability that is often . . . poorly addressed: Ilissa B. Ocko and Steven P. Hamburg, "Climate Impacts of Hydropower: Enormous Differences among Facilities and over Time," *Environmental Science & Technology* 53, 23 (2019): 14070–82.

73   "abysmally low presence . . . wildfires and landslides": Manshi Asher and Prakash Bhandari, "Mitigation or Myth? Impacts of Hydropower Development and Compensatory Afforestation on Forest Ecosystems in the High Himalayas," *Land Use Policy* 100 (January 2021): article 105041.

73   building a dam . . . new *demand* for water: Giuliano Di Baldassarre, interview with author, January 24, 2019.

73   Las Vegas is a textbook case . . . increasing water demand: G. Di Baldassarre et al., "Water Shortages Worsened by Reservoir Effects," *Nature* Sustainability 1 (2018): 617–22.

73   Projected water shortages on the coast of Spain . . . again need more water: J. Pittock, J-H Meng, M. Geiger, and A. K. Chapagain, *Interbasin Water Transfers and Water Scarcity in a Changing World—a Solution or a Pipedream?* (WWF, 2009), https://wwfeu.awsassets.panda.org/downloads/pipedreams18082009.pdf.

73   the reservoir effect: G. Di Baldassarre et al., "Water Shortages Worsened by Reservoir Effects," *Nature Sustainability* 1 (2018): 617–22.

74   Kelsey Leonard: Leonard is at the School of Environment, Resources and Sustainability, University of Waterloo, in Ontario.

74   As international lenders . . . Latin America, Asia, China, and Turkey: Erica Gies, "A Dam Revival, Despite Risks," *New York Times*, November 19, 2014, https://www .nytimes.com/2014/11/20/business/energy-environment/private-funding- brings-a-boom-in-hydropower-with-high-costs.html.

74   more than 3,700 hydropower projects: G. Grill et al., "Mapping the World's Free-Flowing Rivers," *Nature* 569 (2019): 215–21.

74   Cambodia ... put a ten-year hiatus on its planned dams: Stefan Lovgren, "Five Bright Spots on the Mekong," *Circle of Blue*, March 22, 2021, https://www.circleofblue.org/2021/world/five-bright-spots-in-the-mekong/.

75   miles of underground fungal filaments ... communications system for trees: S. Simard et al., "Net Transfer of Carbon between Ectomycorrhizal Tree Species in the Field," *Nature* 388 (1997): 579–82.

75   Systems theory recognizes ... the entire system into dysfunction: G. I. Hagstrom and S. A. Levin, "Marine Ecosystems as Complex Adaptive Systems: Emergent Patterns, Critical Transitions, and Public Goods," *Ecosystems* 20 (2017): 458–76; Paul D. Bakke, Michael Hrachovec, and Katherine D. Lynch, "Hyporheic Process Restoration: Design and Performance of an Engineered Streambed," *Water* 12, no. 425 (2020): 2.

76   some Indigenous people ... species who rely on each other to survive: Karissa Gall, "Iconic Haida Gwaii Species to Be Included in Literary Field Guide for 'Cascadia,'" *Haida Gwaii Observer*, July 2, 2020, https://www.haidagwaiiobserver.com/news/iconic-haida-gwaii-species-to-be-included-in-literary-field-guide-for-cascadia/.

77   Wetlands also store ... that stay wet the longest: A. Nahlik and M. Fennessy, "Carbon Storage in US Wetlands," *Nature Communications* 7 (2016): 13835.

78   In north Seattle on a sunny September day in 2019: Some material in this chapter is also published in a forthcoming article by Erica Gies on the hyporheic zone for *Scientific American*.

78   urban stream syndrome: Christopher J. Walsh et al., "The Urban Stream Syndrome: Current Knowledge and the Search for a Cure," *Journal of the North American Benthological Society* 24, no. 3 (2005): 706–23.

79   flowing downstream but orders of magnitude more slowly: Tim Abbe et al., "Can Wood Placement in Degraded Channel Networks Result in Large-Scale Water Retention?" (Proceedings of the SEDHYD 2019 Conference on Sedimentation and Hydrologic Modeling, June 24–28, 2019, Reno, Nevada), table 1.

79   Along a large river ... the top three to ten feet or so: J. A. Stanford and J. V. Ward, "The Hyporheic Habitat of River Ecosystems," *Nature*, September 1, 1988; Skuyler Herzog, pers. comm., July 12, 2021.

81   Mike Hrachovec: Mike Hrachovec, interview with author, June 15, 2020.

81–82 Because these plants and critters ... without regular flooding: N. Poff et al., "The Natural Flow Regime," *BioScience* 47, no. 11 (1997): 769–84.

82   stream ecologists worldwide ... separating and rejoining: B. Cluer and C. Thorne, "A Stream Evolution Model Integrating Habitat and Ecosystem Benefits," *River Research and Applications* 30 (2014): 135–54.

83   "River management based solely on physical science ... have not been solved.": M. F. Johnson et al., "Biomic River Restoration: A New Focus for River Management," *River Research and Applications* 36 (2020): 3–12.

83   "the power of biology to influence river processes" ... biomic river restoration: Johnson et al., "Biomic River Restoration," 1.

84   "when we wanted to carry ducks ... saved our powder and shot": William Henry Thomes, *On Land and Sea: Or, California in the Years 1843, '44 and '45* (Laird & Lee, 1892), 200.

85   *The Hyporheic Handbook*: See S. Buss et al., *The Hyporheic Handbook: A Handbook*

on the Groundwater–Surface Water Interface and the Hyporheic Zone for Environmental Managers (UK Environment Agency Science Report SCo 50070, 2009).

85   Paul Bakke: Paul Bakke, interview with author, March 9, 2021.

87   They found water mixing at eighty-nine times the preconstruction rate: Bakke, Hrachovec, and Lynch, "Hyporheic Process Restoration."

87   meiofauna, less than a millimeter long: Walter Traunspurger and Nabil Majdi, "Meiofauna," in *Methods in Stream Ecology, Volume 1 (Third Edition)*, ed. F. Richard Hauer and Gary A. Lamberti (Academic Press, 2017), ch. 14, 273–95.

87   nematodes, copepods . . . and tardigrades (a.k.a. water bears): Adrienne Mason, "The Micro Monsters beneath Your Beach Blanket," *Hakai Magazine*, March 21, 2016, https://www.hakaimagazine.com/videos-visuals/micro-monsters-beneath-your-beach-blanket/.

88   Kate Macneale: Kate Macneale, interview with author, July 23, 2020.

89   Urban runoff poisons bugs . . . from vehicle brake pads: Zhenyu Tian et al., "A Ubiquitous Tire Rubber-Derived Chemical Induces Acute Mortality in Coho Salmon," *Science* 371, no. 6525 (2021): 185–89.

90   to seed four creeks . . . sensitive to various stressors: Kate Macneale, *Bug Seeding: A Possible Jump-Start to Stream Recovery* (King County Water and Land Resources Division, 2020).

91   Rhodes and Morley explain their experiment to me: Sarah Morley and Linda Rhodes, interview with author, July 23, 2020.

91–92 The good news is . . . didn't survive long: S. A. Morley et al., "Invertebrate and Microbial Response to Hyporheic Restoration of an Urban Stream," *Water* 13 (2021).

93   Another team of researchers used mass spectrometry . . . the long stretch, 78 percent: Katherine T. Peter et al., "Evaluating Emerging Organic Contaminant Removal in an Engineered Hyporheic Zone Using High Resolution Mass Spectrometry," *Water Research* 150 (2019): 140–52.

94   A geomorphologist with Natural Systems Design: Tim Abbe.

Chapter 4

97   "Beavers mean higher water tables and water on the landscape throughout the dry and wet seasons": Frances Backhouse, interview with author, September 2015.

97   Bison dug or expanded . . . plants can live: Alice Outwater, *Water: A Natural History* (Basic Books, 1997).

98   an estimated sixty million to four hundred million beavers lived on the continent: Kate Lundquist with Brock Dolman, *Beaver in California: Creating a Culture of Stewardship* (Occidental Arts & Ecology Center Water Institute, 2020), https://oaec.org/publications/beaver-in-california/.

98   The longest documented beaver dam . . . twice the length of Hoover Dam: Frances Backhouse, *Once They Were Hats: In Search of the Mighty Beaver* (ECW Press, 2015).

98   "five hundred years ago . . . the rules of engagement": Backhouse, *Once They Were Hats*, 41.

99   Hudson's Bay Company . . . elbow out competing companies: Lundquist and Dolman, *Beaver in California*, 8.

99   Beaver numbers across North America dropped to as few as a hundred thousand: Ben Goldfarb, *Eager: The Surprising, Secret Life of Beavers and Why They Matter*, Chelsea Green, 2019.

99   The animal's decimation fundamentally altered the plumbing of much of North America: Kent Woodruff, a retired biologist with the US Forest Service who was a founder of the modern beaver restoration movement, 2015 interview with author for Erica Gies, "Coca-Cola Leaves It to Beavers to Fight the Drought," *TakePart*, September 23, 2015, http://www.takepart.com/article/2015/09/23/coca-cola-using-beavers-increase-water-supply.

99   in the United Kingdom . . . trees and shrubs: Bryony Coles, *Beavers in Britain's Past* (Oxbow Books, 2006), 59.

100 six of the ten wettest years have occurred since 1998: M. Kendon et al., "State of the UK Climate 2018," *International Journal of Climatology* 39, suppl. 1 (2019): 1–55.

100 Even under a low-impact climate change scenario . . . double: Selma B. Guerreiro et al., "Future Heat-Waves, Droughts and Floods in 571 European Cities," *Environmental Research Letters* 13 (2018).

100 a mimicry that Miwok people . . . create wet microhabitats: Don Hankins (California State University, Chico, professor of geography and planning), "Indigenous Water and Fire Expertise in California," interview, *Water Talk*, season 1, episode 8, June 12, 2020, https://water-talk.squarespace.com/episodes/episode-08.

103 When the project began . . . known locally as Culm: M. Elliott et al., *Beavers— Nature's Water Engineers: A Summary of Initial Findings from the Devon Beaver Projects* (Devon Wildlife Trust, 2017), https://www.devonwildlifetrust.org/sites/default /files/2018-01/Beaver%20Project%20update%20%28LowRes%29%20.pdf.

104 the beavers' expanding numbers . . . cleaning pollution from water: R. E. Brazier et al., *River Otter Beaver Trial: Science and Evidence Report* (University of Exeter, Centre for Resilience, Water and Waste, 2020), https://www.exeter.ac.uk/creww /research/beavertrial/.

104 thirteen dams at Elliot's West Devon site: A. Puttock et al., "Eurasian Beaver Activity Increases Water Storage, Attenuates Flow and Mitigates Diffuse Pollution from Intensively-Managed Grasslands," *Science of the Total Environment* 576 (2017): 430–43, https://doi.org/10.1016/j.scitotenv.2016.10.122.

104–5 six dams on the River Otter . . . reduced peak flood flows: R. E. Brazier et al., *River Otter Beaver Trial: Science and Evidence Report*, https://www.exeter.ac.uk/creww /research/beavertrial/.

105 outflow was three times lower than inflow: A. Puttock et al., "Beaver Dams Attenuate Flow: A Multi-site Study," *Hydrological Processes* 35, no. 2 (2021): article e14017, https://doi.org/10.1002/hyp.14017.

105 "The five-year trial . . . to settle as they need": "Five-Year Beaver Reintroduction Trial Successfully Completed," Gov.UK, August 20, 2020, https://www.gov.uk /government/news/five-year-beaver-reintroduction-trial-successfully-completed.

105 "This was a landmark decision . . . extinct native mammal to England": "Government Landmark Decision Means Beavers Can Stay!," Devon Wildlife Trust, accessed August 2020, https://www.devonwildlifetrust.org/what-we-do/our-projects/river-otter-beaver-trial.

105 Eurasian beavers . . . number 1.2 million across the continent: "Eurasian Beaver: *Castor fiber*," Scottish Wildlife Trust, https://scottishwildlifetrust.org.uk/species /beaver/.

107 beavers and people tend to want to live . . . airports: Ben Dittbrenner (Beavers Northwest), interview with author, September 2019.

107   at two years old ... elsewhere: Alina Bradford, "Facts about Beavers," Live Science, October 13, 2015, https://www.livescience.com/52460-beavers.html.

109   beavers "do much toward ... conservation": Elmo W. Heter, "Transplanting Beavers by Airplane and Parachute," *Journal of Wildlife Management* 14, no. 2 (1950): 143–47.

109   staff distributed 1,221 beavers into watersheds across the state: Lundquist and Dolman, *Beaver in California*, 8.

109   "It is now understood ... upper reaches of many streams": Donald T. Tappe, *The Status of Beavers in California* (Game Bulletin No. 3, California Department of Natural Resources, 1942), https://oaec.org/wp-content/uploads/2016/06/The-Status-Of-Beavers-in-CA.pdf.

110   "California's busy beavers ... fish, wildlife, and agriculture": Lundquist with Dolman, *Beaver in California*, 9.

110   to "resolve wildlife conflicts to allow people and wildlife to coexist": "Wildlife Damage," USDA Animal and Plant Health Inspection Service Wildlife Services, accessed July 19, 2021, https://www.aphis.usda.gov/aphis/ourfocus/wildlifedamage.

110   killing nearly twenty-six thousand beavers in forty-three states: "Program Data Report G—2020: Animals Dispersed / Killed or Euthanized / Removed or Destroyed / Freed or Relocated," United States Department of Agriculture, 2020, https://www.aphis.usda.gov/aphis/ourfocus/wildlifedamage/pdr/?file=PDR-G_Report&p=2020:INDEX:.

110   Ben Dittbrenner: Dittbrenner is now associate professor and director of the Environmental Science and Policy MS program at Northeastern University in Boston.

113   More than a third ... rely on wetlands: "Why Are Wetlands Important?," U.S. Environmental Protection Agency, last updated June 13, 2018, https://www.epa.gov/wetlands/why-are-wetlands-important.

115   To make nests for her eggs ... on a ski slope: S. Buss et al., S. Buss et al., *The Hyporheic Handbook: A Handbook on the Groundwater–Surface Water Interface and the Hyporheic Zone for Environmental Managers* (UK Environment Agency Science Report SC0 50070, 2009).

115   spawning may have lowered ... water flow: Alexander K. Fremier, Brian J. Yanites and Elowyn M. Yager, "Sex That Moves Mountains: The Influence of Spawning Fish on River Profiles over Geologic Timescales," *Geomorphology* 305 (2018): 163–72.

118   the average beaver pond contains ... underground: Heather Simmons, "Beaver Reintroduction a Watershed Success," Washington Department of Ecology, December 2, 2015, https://ecology.wa.gov/Blog/Posts/December-2015/Beaver-reintroduction-a-watershed-success.

118   Beaver water complexes ... climate chaos is wreaking here: E. Fairfax and A. Whittle, "Smokey the Beaver: Beaver-Dammed Riparian Corridors Stay Green during Wildfire throughout the Western USA," *Ecological Applications* 30, no. 8 (2020).

118   the Cascades are projected to lose nearly 100 percent of the snowpack by 2080: P. Mote et al., *Integrated Scenarios of Climate, Hydrology, and Vegetation for the Northwest* (Conservation Biology Institute, 2014).

119   For his doctorate in forest and environmental science: Benjamin J. Dittbrenner, "Restoration Potential of Beaver for Hydrological Resilience in a Changing Climate," PhD diss., School of Environmental and Forest Sciences, University of Washington, 2019.

120   beaver meadows once stored ... about 8 percent: E. Wohl, "Landscape-Scale Car-

bon Storage Associated with Beaver Dams," *Geophysical Research Letters* 40, no. 14 (2013): 3631–36.

120   waxy red snow plants: *Sarcodes sanguinea*, a mycotrophic wildflower that gets its nutrients from underground fungi. "*Sarcodes sanguinea*—Snow Plant," US Forest Service, n.d., https://www.fs.fed.us/wildflowers/beauty/mycotrophic/sarcodes_sanguinea.shtml.

122–23 water levels underground vary with precipitation . . . to sixty-six feet.: S. M. Yarnell et al., *A Demonstration of the Carbon Sequestration and Biodiversity Benefits of Beaver and Beaver Dam Analogue Restoration Techniques in Childs Meadow, Tehama County, California*, Center for Watershed Sciences Technical Report (CWS-2020–01) (University of California, Davis, 2020), 29.

123   They found beavers . . . fifty-two active and abandoned colonies: D. R. Bailey, B. J. Dittbrenner, and K. P. Yocom, "Reintegrating the North American Beaver (*Castor canadensis*) in the Urban Landscape," *WIREs Water* 6 (2019).

Chapter 5

127   From a minivan on the shoulder of Old Mahabalipuram Road: Some material in this chapter first appeared in Erica Gies, "Chennai Ran Out of Water—but That's Only Half the Story," *bioGraphic*, October 30, 2020, https://www.biographic.com/chennai-ran-out-of-water/.

128   Chennai had lost 62 percent of its wetlands between 1980 and 2010: "Land Use Change and Flooding in Chennai," Care Earth Trust, https://careearthtrust.org/flood/.

128   In Southern Africa's Kalahari Desert . . . through the mouth: James G. Workman, *Heart of Dryness: How the Last Bushmen Can Help Us Endure the Coming Age of Permanent Drought* (Walker, 2009).

128   Nabataeans . . . funnel it to cisterns: C. Ortloff, "The Water Supply and Distribution System of the Nabataean City of Petra (Jordan), 300 BC–AD 300," *Cambridge Archaeological Journal* 15, no. 1 (2005): 93–109.

131   reservoir managers were reluctant to release stored water ahead of the monsoon rains: Krupa Ge, *Rivers Remember: #ChennaiRains and the Shocking Truth of a Man-made Flood* (Westland, 2019), 30–32, 75–86, 183–87.

131   The city has seen . . . over the past two decades: Resilient Chennai and Okapi Research & Advisory, *Resilient Chennai City: KALEIDOSCOPE: My City through My Eyes* (Chennai Resilience Centre, 2019), https://resilientchennai.com/strategy/.

131   "till last week . . . What a city!": xquizit (@lexquizit), "#chennairain till last week the residents were booking water tankers and from today they will book rescue boats.What a city!," Twitter, December 1, 2019, https://twitter.com/lexquizit/status/1201138276398223361.

132   "We do not want to compromise . . . degrading the nature": V. Kalaiarasan (Chennai Rivers Restoration Trust, project officer), interview with author, December 2019.

132   Together they produced reports . . . across the entire watershed: "Chennai," *Water as Leverage for Resilient Cities Asia*, Netherlands Enterprise Agency, Office of International Water Affairs, https://english.rvo.nl/subsidies-programmes/water-leverage; "City of 1000 Tanks," City of 1000 Tanks Project, https://www.cityof1000tanks.org/.

133   sixty-one wetlands and water bodies: Jayshree Vencatesan et al., "Comprehensive

Management Plan for Pallikaranai Marsh," Conservation Authority for Pallikaranai Marshland-TNFD and Care Earth Trust, Govt. of Tamil Nadu (2014): 70–82.

134 These *eris* . . . a series of connected depressions: T. M. Mukundan, *The Ery Systems of South India: Traditional Water Harvesting* (Akash Ganga Trust, 2005).

135 he began to research and map them . . . already gone: See Krishnakumar TK, *Indian Columbus* (blog), https://indiancolumbus.blogspot.com/2020/.

135 Joel Pomerantz with his Seep City outings in San Francisco: See Joel Pomerantz, *Seep City*, http://seepcity.org/.

135–36 British engineers . . . a pattern that continued after independence: Mukundan, *Ery Systems of South India*, 14.

136 Fewer than one-third of the 650 water bodies . . . remain: The 2019 report from Resilient Chennai states that "almost 3,482 water bodies" were "transferred to built-up areas" early in Chennai's colonial development period so there may have been significantly more than the number KK has found. Resilient Chennai and Okapi Research & Advisory, *Resilient Chennai City: KALEIDOSCOPE*, 29.

136 one-fifth its 1893 extent: K. Lakshmi, "The Vanishing Waterbodies of Chennai," *Hindu*, April 1, 2018, https://www.thehindu.com/news/cities/chennai/the-vanishing-waterbodies-of-chennai/article23404437.ece.

139 The government has also begun . . . the Adyar River: "Corporation Reclaims Chennai Water Ways, 90% Encroachers along Cooum Relocated," *Times of India*, November 3, 2020, https://timesofindia.indiatimes.com/city/chennai/corpn-reclaims-city-water-ways-90-encroachers-along-cooum-relocated/articleshow/79007606.cms.

139 proximity relocation: Krishna Mohan (Resilient Chennai, chief resilience officer), interview with author, November 2019.

140 Sudhee NK: Sudheendra Krisnhamurty.

140 Naaz Gani: Naaz Gani, interview with author, December 2019

141 It's not just poor people who lack city water connections: Ashok Natarajan (Tamilnadu Water Investment Company, a private partner of the city's water utility, then CEO, now retired), interview with author, December 4, 2019.

141 Almost fifty-three million gallons daily are shipped around the city: Natarajan, interview, December 4, 2019.

141 Water served up in tanker trucks is often extracted from underground: Lakhshmanan Venkatachalam, "Informal Water Markets and Willingness to Pay for Water: A Case Study of the Urban Poor in Chennai City, India," *International Journal of Water Resources Development* 31, no. 1 (2015): 134–45.

141 Chennai's groundwater table is dropping about four to eight inches every year: *City of 1000 Tanks* (Water as Leverage for Resilient Cities: Asia, Phase 2 Report, May 15, 2019, 2019), 45, https://waterasleverage.org/file/download/57980072/CITY-OF-1,000-TANKS.pdf.

145 Replicating the project . . . 6 percent of the city's demand: *City of 1000 Tanks*, 10.

146 A year after my visit . . . for scientists to study its ecology: P. Oppili, "Thazhambur Lake, Ravaged by Quarrying, Restored," *Times of India*, December 16, 2020, https://timesofindia.indiatimes.com/city/chennai/thazhambur-lake-ravaged-by-quarrying-restored/articleshow/79750535.cms.

148 bringing the total to 358 acres: K. Ilangovan, interview with author, December 3, 2019, and follow-up.

149 several of the projects . . . under consideration by the central government in Delhi: Current as of May 2021, according to Vencatesan.

149   A state wetlands panel recently . . . another thirty-one: K. Lakshmi, "Four Chennai Lakes Will Soon Be Notified as Wetlands to Help Conserve Them," *The Hindu,* October 22, 2021, https://www.thehindu.com/news/cities/chennai/four-chennai-lakes-will-soon-be-notified-as-wetlands/article37124295.ece.

Chapter 6

150   On a mild day pre-pandemic: Some of this chapter was first published from reporting for Erica Gies, "Why Peru Is Reviving a Pre-Incan Technology for Water," *BBC Future,* May 18, 2021, https://www.bbc.com/future/article/20210510-perus-urgent-search-for-slow-water.

150   "summer every day to winter every night.": The German geographer Karl Troll once described the climate in the high Andes as "winter every night and summer every day."

151   Canal irrigation in northern Peru's Zaña Valley dates back to 4700 BCE: Tom D. Dillehay, Herbert H. Eling, and Jack Rossen, "Preceramic Irrigation Canals in the Peruvian Andes." *Proceedings of the National Academy of Sciences* 102, no. 47 (2005): 17241–44.

151   Peru is seeing water scarcity grow worse as a result of climate change and human activities: Noah Walker-Crawford and Angela Thür, "Dying Slower in a Changing Climate: Water Scarcity and Flood Hazard in the Peruvian Andes," *PLOS Collections* (blog), June 14, 2019, https://collectionsblog.plos.org/dying-slower-in-a-changing-climate-water-scarcity-and-flood-hazard-in-the-peruvian-andes/.

151   In Nazca . . . push water along like a pump: Rosa Lasaponara, Nicola Masini, and Giuseppe Orefici, eds., *The Ancient Nasca World: New Insights from Science and Archaeology* (Springer, 2016).

151–52 In its heyday . . . *puquios* are still used today: Katharina J. Schreiber and Josué Lancho Rojas, "The Puquios of Nasca," *Latin American Antiquity* 6, no. 3 (1995): 229–54.

152   *Stolen Continents*: See Ronald Wright, *Stolen Continents: Five Hundred Years of Conquest and Resistance in the Americas* (Houghton Mifflin Harcourt, 2005).

152   *The First New Chronicle and Good Government*: Felipe Guamán Poma De Ayala, *The First New Chronicle and Good Government* (University of Texas Press, 2006).

153–54 around 1.5 billion people could depend . . . in the 1960s: Daniel Viviroli et al., "Increasing Dependence of Lowland Populations on Mountain Water Resources," *Nature Sustainability* 3, no. 11 (2020): 917–28.

154   1.5 million people who are not connected to city water: Natalie Jean Burg et al., "Access to Water for Human Consumption in Lima, Peru: An Analysis of Challenges and Solutions" (online presentation, preconference event, Second International Conference, Water, Megacities and Global Change, December 2020).

154   Lima's water management will be inadequate as early as 2030: David G. Groves et al., *Preparing for Future Droughts in Lima, Peru: Enhancing Lima's Drought Management Plan to Meet Future Challenges* (World Bank, 2019).

157   Paiute canals to recharge aquifers in Southern California: Don Hankins (California State University, Chico, professor of geography and planning), interview, *Water Talk,* "Indigenous Water and Fire Expertise in California," season 1, episode 8, June 12, 2020. https://water-talk.squarespace.com/episodes/episode-08.

160 these societies . . . "were built on the ethic of reciprocity, not rapacity": Wright, *Stolen Continents*, 73.

165 the findings of their study . . . water scarcity in Peru: See Boris R. Ochoa-Tocachi et al., "Potential Contributions of Pre-Inca Infiltration Infrastructure to Andean Water Security," *Nature Sustainability* 2, no. 7 (2019): 584–93.

166 Nevertheless, what remained . . . would be even higher: Some material in this story was adapted from Erica Gies, "Seeking Relief from Dry Spells, Peru's Capital Looks to Its Ancient Past," *National Geographic*, July 9, 2019, https://www .nationalgeographic.com/environment/article/seeking-relief-from-drought-peru-capital-lima-looks-to-ancient-past.

166–67 That's a barrier to . . . the building of additional projects: Sophie Tremolet (Nature Conservancy in Europe, a water economist; formerly with the World Bank), interview with author, April 2020.

167 SEDAPAL is investing . . . conserving water for crops and reducing erosion: Oscar Angulo (Forest Trends, water and sanitation coordinator for natural infrastructure investment), interview with author, April 2021.

168 Global trade tends to exacerbate . . . rein in consumption: Oliver Taherzadeh, Mike Bithell, and Keith Richards, "Water, Energy and Land Insecurity in Global Supply Chains," *Global Environmental Change* 67 (March 2021): article 102158.

169 "The ecosystems on which life itself is based . . . how water is managed today": "Report Warns 700m People at Risk of Displacement by Intense Water Scarcity by 2030," Water Briefing Global, March 14, 2018, https://www.waterbriefingglobal .org/report-warns-700m-people-at-risk-of-displacement-by-intense-water-scarcity-by-2030/. See also *Making Every Drop Count: An Agenda for Water Action* (UN High Level Panel on Water, 2018), https://sustainabledevelopment.un.org /content/documents/17825HLPW_Outcome.pdf.

169 "More than 34 countries" . . . and it expected many more to join: "UN Adopts Landmark Framework to Integrate Natural Capital in Economic Reporting," United Nations, March 2021, https://www.un.org/en/desa/un-adopts-landmark-framework-integrate-natural-capital-economic-reporting.

169 how much it would cost . . . to manufacture replacements: Robert Costanza et al., "The Value of the World's Ecosystem Services and Natural Capital," *Nature* 387, no. 6630 (1997): 253–60.

170 $125 trillion annually, which was then more than one-and-a-half times the global GDP: Robert Costanza et al., "Changes in the Global Value of Ecosystem Services," *Global Environmental Change* 26 (2014): 152–58.

170 With these values in mind . . . inside the system: James Salzman et al., "The Global Status and Trends of Payments for Ecosystem Services," *Nature Sustainability* 1, no. 3 (2018): 136–44.

170 a letter published in the journal *BioScience* in 2019: William J. Ripple, et al., "World scientists' warning of climate emergency," *BioScience* 70, no. 1 (2019): 8–12.

172 forty of the country's fifty water utilities . . . that holds water: Oscar Angulo (Forest Trends, water and sanitation coordinator for natural infrastructure investment), interview and emails with the author, April–June 2021.

172 peatlands . . . store 10 percent of all fresh water: Fereidoun Rezanezhad et al., "Structure of peat soils and implications for water storage, flow and solute transport: A review update for geochemists," *Chemical Geology* 429 (2016): 75–84.

172 biodiversity hotspots . . . domesticated alpacas and llamas: M. S. Maldonado

Fonkén, "An introduction to the bofedales of the Peruvian High Andes," *Mires and Peat* 15 (2014): 1–13.

173   "This is a public service announcement . . . Use compost instead": Cecilia Gianella, pers. comm., July 24, 2019.

173   Officials from SEDAPAL . . . the healthy *bofedales* that remain: Angulo, interview, April 2021.

175   The guidelines will instruct local people about what to do: Angulo, interview, April 2021.

177   a 2020 analysis . . . greater hydrological benefits: Vivien Bonnesoeur et al., "Impacts of Infiltration Trenches on Hydrological Ecosystem Services: A Systematic Review" (American Geophysical Union Fall Meeting, December 15, 2020, online).

Chapter 7

181   Around the world . . . costs rising from $17 billion to $177 billion: Samantha Kuzma and Tianyi Luo, "The Number of People Affected by Floods Will Double between 2010 and 2030," World Resources Institute, April 23, 2020, https://www.wri.org/blog/2020/04/aqueduct-floods-investment-green-gray-infrastructure.

182   the Mississippi alone has 3,500 miles of levees averaging twenty-five feet high: Jacoby Smith and Chongzi Wang, "Mississippi River Basin," ArcGIS StoryMaps, November 1, 2019, https://storymaps.arcgis.com/stories/81c42fcedf084856b237fb12a60d00ee.

182   The US General Accounting Office . . . over the twentieth century: Nicholas Pinter, "One Step Forward, Two Steps Back on U.S. Floodplains," *Science* 308, no. 5719 (2005): 207–8, at 208.

182   the Army Corps of Engineers . . . "incremental loss of floodplain land to development": Pinter, "One Step Forward, Two Steps Back," 207.

182   levees also multiply flood damage . . . overtop or fail: E. Ridolfi, F. Albrecht, and G. Di Baldassarre, "Exploring the Role of Risk Perception in Influencing Flood Losses over Time," *Hydrological Sciences Journal* 65, no. 1 (2019).

183   "Floods are acts of god," he writes, "but flood losses are largely acts of man": Gilbert F. White, "Human Adjustments to Floods" (Research Paper 29, Department of Geography Research, University of Chicago, 1945).

183   a community's experience of a major flood . . . within a decade: A. Viglione et al., "Insights from Socio-hydrology Modelling on Dealing with Flood Risk—Roles of Collective Memory, Risk-Taking Attitude and Trust," *Journal of Hydrology* 518, part A (2014): 71–82.

183   the historic flood in 1993 . . . land that had been inundated: Pinter, "One Step Forward, Two Steps Back."

183   Charles Ellet argued for a multipronged approach to mitigate flooding: Charles Ellet, *Report on the Overflows of the Delta of the Mississippi* (AB Hamilton, 1852).

183   A. A. Humphreys . . . a "levees only" approach: Andrew Atkinson Humphreys and Henry L. Abbot, *Report upon the physics and hydraulics of the Mississippi river: upon the protection of the alluvial region against overflow; and upon the deepening of the mouths . . . Submitted to the Bureau of Topographical Engineers. War Department, 1861. No. 4.* (J. B. Lippincott, 1861).

184   "river-training structures" . . . raise flood levels many feet: Nicholas Pinter (UC

Davis, hydrogeologist and associate director of the Center for Watershed Sciences) interview and emails with author, 2020–21. Pinter, who studies flooding, previously taught at the University of Illinois at Urbana-Champaign.

184 Part of what's now the Corn Belt . . . funnel away excess water: Sharon Levy, "Learning to Love the Great Black Swamp," *Undark*, March 31, 2017. https://undark .org/2017/03/31/great-black-swamp-ohio-toledo/.

184 Corn Belt states . . . banished 85 to 90 percent of their native wetlands : Thomas E. Dahl, *Wetlands Losses in the United States, 1780's to 1980's* (US Department of the Interior, Fish and Wildlife Service, 1990), 6, table 1, https://www.fws.gov/wetlands /documents/Wetlands-Losses-in-the-United-States-1780s-to-1980s.pdf.

184 they were sued by 372 property owners: *Ideker Farms, Inc. v. United States*, 136 Fed. Cl. 654 (2018).

184 "military hydrological complex projects": Pinter, interview, July 17, 2020.

185 Olivia Dorothy: Olivia Dorothy, interview with author, October 2, 2020.

185 Marsh managers . . . are looking for more: Brian Ritter (Nahant Marsh, executive director; Eastern Iowa Community Colleges, conservation program coordinator), email with the author, May 28, 2021.

185 "let floodplains be floodplains": Craig Just, University of Iowa professor and Iowa Flood Center researcher, quoted in Rocky Kistner, "Thinking Outside the Box: How Davenport Uses Marshes to Combat Floods and Climate Change," *How We Respond*, American Association for the Advancement of Science, 2019, https:// howwerespond.aaas.org/community-spotlight/thinking-outside-the-box-how-davenport-uses-marshes-to-combat-floods-and-climate-change/.

186 Restoring wetlands . . . by as much as 29 percent: Meghna Babbar-Sebens et al., "Spatial Identification and Optimization of Upland Wetlands in Agricultural Watersheds," *Ecological Engineering* 52 (2013): 130–42.

186 Henk Ovink: Henk Ovink, interview with author, August 2015.

186 a strong storm blew out their dikes: Cynthia Barnett, *Blue Revolution: Unmaking America's Water Crisis* (Beacon Press, 2011), 47–49.

187 After 1995, the government decided it needed a different approach: Some material in the section was derived from Erica Gies, "Cities Are Finally Treating Water as a Resource, Not a Nuisance," *Ensia*, September 1, 2015, https://ensia.com/features /cities-are-finally-treating-water-as-a-resource-not-a-nuisance/.

188 62 percent of China's cities flooded: Hui Li et al., "Sponge City Construction in China: A Survey of the Challenges and Opportunities," *Water* 9, no. 9 (2017): 594.

188 In 1960, fewer than 34 percent . . . to more than four billion today: *World Urbanization Prospects: The 2018 Revision* (United Nations Department of Economic and Social Affairs/Population Division, 2018), https://population.un.org/wup /Publications/Files/WUP2018-Report.pdf.

188 the land area covered by cities . . . since 1992: John Bongaarts, "IPBES, 2019: Summary for Policymakers of the Global Assessment Report on Biodiversity and Ecosystem Services of the Intergovernmental Science–Policy Platform on Biodiversity and Ecosystem Services," 2019, 680–81.

188 Greater New York occupies 4,669 square miles . . . Johannesburg–Pretoria, 1,560: *Demographia World Urban Areas, 17th Annual Edition* (Demographia, June 2021), http://demographia.com/db-worldua.pdf.

188 every time a city increases . . . nearby waterways by 3.3 percent: Annalise G. Blum et

al., "Causal Effect of Impervious Cover on Annual Flood Magnitude for the United States," *Geophysical Research Letters* 47, no. 5 (2020).

189   Up to eighteen inches of rain chucked down on greater Beijing: Parts of this chapter from this point are adapted and updated from Erica Gies, "Sponge City Revolution," *Scientific American*, December 2018.

189–90   The city is lowering . . . causing the ground to sink as well: Dong Jiang, Gang Liu, and Yongping Wei, "Monitoring and Modeling Terrestrial Ecosystems' Response to Climate Change 2016," *Advances in Meteorology*, January 2016, article 5984595, 165–82.

190–91   The Ministry of Housing . . . would run 86.5 billion RMB (US$13.5 billion): Chris Zevenbergen, Dafang Fu, and Assela Pathirana, "Transitioning to Sponge Cities: Challenges and Opportunities to Address Urban Water Problems in China," *Water* 10, no. 9 (2018): 1230.

191   Difficulty accessing data . . . and evaluation standards: Chris Zevenbergen, interview with author, October 2, 2020.

191   The goal was for 20 percent of cities . . . to meet the goal by 2030: United Nations, *The World's Cities in 2018: Data Booklet* (Department of Economic and Social Affairs, Population Division, 2018), "Annex Table," 13–16.

192   Zhengzhou . . . to avert disaster: Steven Lee Myers, Keith Bradsher, and Chris Buckley, "As China Boomed, It Didn't Take Climate Change into Account. Now It Must," *New York Times*, July 26, 2021.

192   Niall Kirkwood: Niall Kirkwood, interview with author, September 6, 2018.

194   Jack Dangermond: Jack Dangermond, interview with author, May 20, 2019.

195   Ryan Perkl: Ryan Perkl, interview with author, June 14, 2018.

195   If soil at a project site . . . mix in sand: Erica Gies, "As Floods Increase, Cities Like Detroit Are Looking to Green Stormwater Infrastructure," *Ensia*, April 16, 2019, https://ensia.com/features/flooding-increase-cities-live-with-water-green-stormwater-infrastructure/.

196   When their landscape maps are complete . . . stormwater behavior: Perkl, interview, June 14, 2018.

197   Yanweizhou Park . . . using only biological processes: Erica Gies, "Sponge City Revolution," *Scientific American*, December 2018, https://ericagies.com/wp-content/uploads/2020/04/Sponge-City-Revolution-Gies-SciAm.pdf.

199   In the United States, many are replaced after fifty years: Robin Grossinger (San Francisco Estuary Institute, historical ecologist), interview with author, November 16, 2018.

199   In rapid-growth places like China, that turnover may happen in just fifteen years: Randy Dahlgren (soil scientist at U.C. Davis who has worked extensively in China), interview with author, June 14, 2018. Dahlgren's insights are based on his personal observations since he started traveling to China in 2002. Given the lack of private land ownership, the government can develop infrastructure projects rapidly. For more, see Jingjing Wang, Yurong Zhang and Yuanfeng Wang, "Environmental Impacts of Short Building Lifespans in China Considering Time Value," *Journal of Cleaner Production* 203 (2018): 696–707.

199   The parks opened in May 2018 . . . the formerly deserted waterfront: "Revitalizing Kazan's Prime Waterfront, Russia," World Architects, https://www.world-architects.com/en/turenscape-haidian-district-beijing/project/revitalizing-kazan-s-prime-waterfront-russia.

201 similar concern about a cookie-cutter approach: Hui Li et al., "Sponge City Construction in China," 594.

202 killed at least 219 people: Steven Lee Myers, "After Covid, China's Leaders Face
New Challenges from Flooding," *New York Times*, August 21, 2020, https://www.
nytimes.com/2020/08/21/world/asia/china-flooding-sichuan-chongqing.html.

202 333 rivers have dried up: Announcement No. 3 of 2018: Audit Results of Ecological
Environment Protection in the Yangtze River Economic Zone, http://www.audit.
gov.cn/n5/n25/c123511/content.html.

203 Randy Dahlgren: Randy Dahlgren, interview with author, June 14, 2018.

204 As of June 2021 ... two billion gallons annually: Stephanie Chiorean (environmental staff scientist and planner, Philadelphia Water Department), email with the
author, June 14, 2021.

204 bacteria in soil ... increased serotonin levels: Christopher A. Lowry et al., "Identification of an Immune-Responsive Mesolimbocortical Serotonergic System: Potential Role in Regulation of Emotional Behavior," *Neuroscience* 146, no. 2 (2007):
756–72.

204 Compounds emitted by trees boost the immune system and provide relaxation:
Andrew Nikiforuk, "How a Famous Tree Scientist Seeks Well-Being in Nature
during the Pandemic," *Tyee*, December 31, 2020, https://thetyee.ca
/News/2020/12/31/Tree-Scientist-Seeks-Nature-Well-Being-Pandemic/.

205 we have remarkably similar plant species in cities the world over: E. Spotswood et
al., *Re-oaking Silicon Valley: Building Vibrant Cities with Nature*, SFEI Contribution
No. 825 (San Francisco Estuary Institute, 2017), http://www.sfei.org/documents
/re-oaking-silicon-valley.

205 The largest irrigated crop ... nine billion gallons of water per day: "Outdoor Water
Use in the United States," US Environmental Protection Agency, last updated February 14, 2017, https://19january2017snapshot.epa.gov/www3/watersense/pubs
/outdoor.html.

205–6 "When we try to make something ... outside those refugia": Maleea Acker, "Why
Are We in Trouble?," *Focus on Victoria*, November 2019. https://www.focusonvicto-
ria.ca/commentary/why-are-we-in-trouble-r33/.

206 Human health, too, benefits from increasing biodiversity in our cities: Jacob G.
Mills et al., "Relating Urban Biodiversity to Human Health with the 'Holobiont'
Concept," *Frontiers in Microbiology* 10 (2019): 550.8.

207 *Biidaaban: First Light*: Lisa Jackson, *Biidaaban: First Light VR*, 2018, http://
lisajackson.ca/Biidaaban-First-Light-VR.

Chapter 8

210 A single tree can transpire ... farther away: David Ellison et al., "Trees, Forests and
Water: Cool Insights for a Hot World," *Global Environmental Change* 43 (2017): 51–
61.

210 Wangari Maathai: Wangari Maathai, *Unbowed, A Memoir* (Anchor Books, 2006).

210–11 Kenya gets 34 percent ... unreliable: Emilio F. Moran et al., "Sustainable Hydropower in the 21st Century," *Proceedings of the National Academy of Sciences* 115, no. 47
(2018): 11891–98.

211 But there's also a global phenomenon ... hand it over to corporations: Erica Gies,

"Investors Are Grabbing a Japan-Size Chunk of the Developing World for Food and Water," *TakePart*, August 28, 2015, http://www.takepart.com/article/2015/08/28 /land-grabs-secure-water-rich-countries-cost-poor.

211   1,350 square miles in Kenya: "Land Matrix," Land Matrix Initiative, https:// landmatrix.org/map.

212   the continent is projected to warm up to 1.5 times more than the global average: "Global Warming: Severe Consequences for Africa," Africa Renewal, March 2019, https://www.un.org/africarenewal/magazine/december-2018-march-2019 /global-warming-severe-consequences-africa.

212   the number of girls reaching secondary education . . . from 7.6 to 4: Daphne H. Liu and Adrian E. Raftery, "How Do Education and Family Planning Accelerate Fertility Decline?," *Population and Development Review*, July 23, 2020.

212   40 percent of women in sub-Saharan Africa still lack access to reproductive health care: C. Pons et al., *Inequalities in Women's and Girls' Health Opportunities and Outcomes: A Report from Sub-Saharan Africa* (World Bank Group, 2016), https://www .isglobal.org/documents/10179/5808952/Report+Africa.pdf.

216   New York City famously bought . . . $250 million a year for maintenance: *Watershed Management for Potable Water Supply: Assessing the New York City Strategy* (National Research Council, National Academies Press, 2000).

217   Only 53 percent of people in the capital have direct access to water: "Providing Sustainable Sanitation and Water Services to Low-income Communities in Nairobi," World Bank, February 19, 2020, https://www.worldbank.org/en/news /feature/2020/02/19/providing-sustainable-sanitation-and-water-services-to-low-income-communities-in-nairobi.

217   those who buy water with e-money . . . city water rates: "Kenya's Informal Settlements Need Safe Water to Survive Covid-19," United Nations Human Rights, April 6, 2020, https://www.ohchr.org/EN/NewsEvents/Pages/COVID19_ RighttoWaterKenya.aspx.

217   TNC has raised more than $20 million: "The Value of Water: Making a Business Case for One of Kenya's Most Vital Resources," Resilient Food Systems, April 16, 2021, https://resilientfoodsystems.co/news/the-value-of-water-making-a-business-case-for-one-of-kenyas-most-vital-resources.

219   So far, improved soil management . . . compared with 2016 levels: Craig Leisher, pers. comm., June 11, 2021.

222   this area was native forest during her childhood in the 1940s and '50s: Maathai, *Unbowed*, 38.

223   Craig Leisher: Craig Leisher, interview with author, March 31, 2020.

226   In recent years, the Green Belt Movement . . . underground water flow: T. Paul Cox, "Watersheds: A Common Destiny for Survival," *New Agriculturalist*, January 2012, http://www.new-ag.info/en/focus/focusItem.php?a=2363.

226–27 Burning plants for these uses . . . highest rates of deforestation: *Africa Energy Outlook 2019* (International Energy Agency, November 2019), https://www.iea.org /reports/africa-energy-outlook-2019.

227   Kenya also has a long history of government officials giving away public lands to cronies: Maathai, *Unbowed*, 255, 261, 281.

227   government officials evicted . . . in northwest Kenya: "Kenya: Sengwer Evictions from Embobut Forest Flawed and Illegal," Amnesty International, May 15, 2018,

https://www.amnesty.org/en/latest/news/2018/05/kenya-sengwer-evictions-from-embobut-forest-flawed-and-illegal/.

227 nature is healthier ... conservation by governments: Brondizio et al., *Global Assessment Report on Biodiversity*; Richard Schuster et al., "Vertebrate Biodiversity on Indigenous-Managed Lands in Australia, Brazil, and Canada Equals that in Protected Areas," *Environmental Science & Policy* 101 (2019): 1–6.

228 Mutundu trees: *Croton macrostachyus*. It has medicinal uses.

229 People believe it has poisonous spittle ... send bad spirits your way: Credo Mutwa, a Southern African Sangomas, or witch doctor, transcribed in "Reptiles—African Folklore" (module 5, component 3, Trees, Reptiles and the Natural World—African Folklore Course, Wildlife Campus, n.d.), http://www.wildlifecampus.com/Courses/AfricanFolklorebyCredoMutwa/TreesReptilesandtheNaturalWorld/Reptiles/220.pdf.

230 China was ahead of the curve ... since 1978: Mark Zastrow, "China's Tree-Planting Could Falter in a Warming World," *Nature*, September 23, 2019, https://media.nature.com/original/magazine-assets/d41586-019-02789-w/d41586-019-02789-w.pdf.

231 Intact, healthy forests ... planted monoculture: E. Dinerstein et al., "A Global Deal for Nature: Guiding Principles, Milestones, and Targets," *Science Advances* 5, no. 4 (2019).

231 half the rain in the Amazon rainforest came from the trees themselves: Eneas Salati et al., "Recycling of Water in the Amazon Basin: An Isotopic Study," *Water Resources Research* 15, no. 5 (1979): 1250–58.

231 Trees can also generate rain ... in the Balkans: Roni Avissar and David Werth, "Global Hydroclimatological Teleconnections Resulting from Tropical Deforestation," *Journal of Hydrometeorology*, April 1, 2005.

231 deforestation can reduce rainfall magnitude: Confidence Duku and Lars Hein, "The Impact of Deforestation on Rainfall in Africa: A Data-Driven Assessment," *Environmental Research Letters* 16 (2021).

231 the biotic pump: Anastassia M. Makarieva et al., "Where Do Winds Come From? A New Theory on How Water Vapor Condensation Influences Atmospheric Pressure and Dynamics," *Atmospheric Chemistry and Physics* 13, no. 2 (2013): 1039–56.

231 It might also explain ... an arid interior: Fred Pearce, "Weather Makers," *Science*, June 19, 2020, https://science.sciencemag.org/content/368/6497/1302.summary.

232 Douglas Sheil: Douglas Sheil, interview with author, November 12, 2020.

232 The researchers looked at ... pine plantations: R. B. Jackson et al., "Trading Water for Carbon with Biological Sequestration," *Science*, December 23, 2005.

232 In California's northern mountains ... scarce water supplies: Will Parrish, "Logging for Water," *Monthly*, August 1, 2016, http://www.themonthly.com/feature1608.html.

232 the loss of so many trees ... recharging groundwater: J. A. Biederman et al., "Recent Tree Die-Off Has Little Effect on Streamflow in Contrast to Expected Increases from Historical Studies," *Water Resources Research* 51 (2015): 9775–89.

Chapter 9

238–39 They can reduce by half ... sea-level rise: K. Arkema et al., "Coastal Habitats Shield People and Property from Sea-Level Rise and Storms," *Nature Climate Change* 3 (2013): 913–18.

239 It's a golden summer day in 2018: Some material in this chapter has been adapted from Erica Gies, "Fortresses of Mud: How to Protect San Francisco Bay from the Rising Seas," *Nature*, October 9, 2018, https://www.nature.com/articles/d41586-018-06955-4.

240 Coastal ecosystems ... three times more carbon per acre than tropical forests: IUCN, "Issues Brief: Blue Carbon" (International Union for Conservation of Nature, November 2017), https://www.iucn.org/sites/dev/files/blue_carbon_issues_brief.pdf. Note that inland wetlands also sequester carbon. In fact, a 2016 study found that inland wetlands in the continental US are so extensive that they hold nearly ten times more stored carbon than those on the coasts. A. Nahlik and M. Fennessy, "Carbon Storage in US Wetlands," *Nature Communications* 7 (2016): article 13835.

240 when faced with sea-level rise ... more carbon dioxide in their soils: Kerrylee Rogers et al., "Wetland Carbon Storage Controlled by Millennial-Scale Variation in Relative Sea-Level Rise," *Nature* 567, no. 7746 (2019): 91–95.

240 Over the past half century ... undermining coastal resilience: Andy Steven et al., "Coastal Development: Resilience, Restoration and Infrastructure Requirements," World Resources Institute, 2020, www.oceanpanel.org/blue-papers /coastaldevelopment-resilience-restoration-and-infrastructure-requirements.

240 about 1,313–3,784 square miles ... stored carbon into the atmosphere: Michael Oppenheimer et al., "Sea Level Rise and Implications for Low Lying islands, coasts and communities," in *Special Report on the Ocean and Cryosphere in a Changing Climate* (Intergovernmental Panel on Climate Change, 2019).

241 salt water is moving into the delta at an unprecedented rate: Ho Huu Loc et al., "Intensifying Saline Water Intrusion and Drought in the Mekong Delta: From Physical Evidence to Policy Outlooks," *Science of The Total Environment* 757 (2021).

241 Sea levels have risen ... since 1993: "How Long Have Sea Levels Been Rising? How Does Recent Sea Level Rise Compare to That over the Previous Centuries?," NASA Earth Data, https://sealevel.nasa.gov/faq/13/how-long-have-sea-levels-been-rising-how-does-recent-sea-level-rise-compare-to-that-over-the-previous/.

242 They are also traveling ... rains linger longer: J. P. Kossin, "A Global Slowdown of Tropical-Cyclone Translation Speed," *Nature* 558 (2018): 104–7.

242 the frequency of sunny-day flooding has increased by five times in some US cities: William Sweet et al., *2019 State of U.S. High Tide Flooding with a 2020 Outlook* (NOAA Technical Report NOS CO-OPS 092, National Oceanic and Atmospheric Administration, July 2020), https://tidesandcurrents.noaa.gov/publications /Techrpt_092_2019_State_of_US_High_Tide_Flooding_with_a_2020_Outlook_30June2020.pdf.

242–43 Seawater can infiltrate ... flood homes from below: K. M. Befus et al., "Increasing Threat of Coastal Groundwater Hazards from Sea-Level Rise in California," *Nature Climate Change* 10 (2020): 946–52.

244 John Bourgeois: John Bourgeois is now deputy officer of stewardship and planning at the Santa Clara Valley Water District.

245 Sea-level rise is expected to accelerate ... by as much as seven feet by 2100: G. Griggs et al., *Rising Seas in California: An Update on Sea-Level Rise Science* (California Ocean Science Trust, April 2017).

245 a vast, loose group of scientists ... the bay had historically: M. Monroe et al., *Bay-*

lands Ecosystem Habitat Goals (US Environmental Protection Agency and S.F. Bay Regional Water Quality Control Board, 1999), 328.

245   about 28 percent of the historic marsh area . . . restored to marsh: "State of the Estuary, 2019 Update," San Francisco Estuary Partnership, 2019, https://www.sfestuary .org/our-estuary/soter/.

248   the traveler's account of San Francisco: William H. Thomes, *On Land and Sea* (Laird & Lee,1892).

248   "At high water . . . as builders afterward discovered": Thomes, *On Land and Sea*, 185.

248   "The United States made short work . . . a wide berth": Thomes, *On Land and Sea*, 189.

250   Without supplemental sediment, many will drown by 2100: S. Dusterhoff et al., *Sediment for Survival: A Strategy for the Resilience of Bay Wetlands in the Lower San Francisco Estuary* (San Francisco Estuary Institute, 2021).

250   Dave Halsing: Dave Halsing, interview with author, December 3, 2020.

251   If sea levels rise by 6.9 feet . . . six hundred million tons of sediment: Dusterhoff et al., *Sediment for Survival*.

252   At the peak, hydraulic miners . . . toward the bay: Scott A. Wright and David H. Schoellhamer, "Trends in the Sediment Yield of the Sacramento River, California, 1957–2001," *San Francisco Estuary and Watershed Science* 2, no. 2 (2004).

252   so much dirt . . . on the bottom of San Francisco Bay: Powell Greenland, *Hydraulic Mining in California: A Tarnished Legacy* (A. H. Clark, 2001), 244.

252   diminishing the size of the bay by one-third: "Prevented Bay Fill," Save the Bay, https://savesfbay.org/impact/prevented-bay-fill.

255   a state senator secured $31.4 million in funding for the project: California State Senator Bob Wieckowski, "Wieckowski to Present Check for Alameda Creek Project," press release, September 25, 2019, https://sd10.senate.ca.gov/news/2019-09-25- wieckowski-present-check-alameda-creek-restoration-project.

256   during the dry season . . . too salty to farm: Doan Van Binh et al., "Long-Term Alterations of Flow Regimes of the Mekong River and Adaptation Strategies for the Vietnamese Mekong Delta," *Journal of Hydrology: Regional Studies* 32 (2020).

256   much of the delta could be inundated . . . as early as 2050: S. A. Kulp and B. H. Strauss, "New Elevation Data Triple Estimates of Global Vulnerability to Sea-Level Rise and Coastal Flooding," *Nature Communications* 10, no. 4844 (2019).

257   more dams are planned . . . on its tributaries: In March 2020, Cambodia announced it would postpone moving forward on its two (of the total eleven) planned mainstem dams until after 2030.

257   existing dams and those under construction would trap . . . before reaching the delta: G. M. Kondolf, Z. K. Rubin, and J. T. Minear, "Dams on the Mekong: Cumulative Sediment Starvation," *Water Resources Research* 50 (2014): 5158–69.

257   What we do on land really matters . . . sea level rise: Sepehr Eslami et al., "Projections of Salt Intrusion in a Mega-delta under Climatic and Anthropogenic Stressors," *Communications Earth and Environment*, July 15, 2021, https://www.nature .com/articles/s43247-021-00208-5.

257   The Mekong Dam Monitor . . . via satellite: "Mekong Dam Monitor," Stimson Eyes on Earth, https://monitor.mekongwater.org/.

257   That has required . . . more than 12,400 miles of levees: Chu Thai Hoanh, Diana Suhardiman, and L. A. Tuan, "Irrigation Development for Rice Production in the

Mekong Delta, Vietnam: What's Next?" (paper presented at the 28th International Rice Research Conference, Hanoi, November 8–11, 2010).

257 overall subsidence . . . dropping by two inches: O. Neusser, *Trouble Underground—Land Subsidence in the Mekong Delta* (Deutsche Gesellschaft für Internationale Zusammenarbeit [GIZ] GmbH, 2019).

258 Now people in the delta are first considering . . . before deciding on land use: SFEI is working on a similar concept it calls "nature's jurisdictions," divvying up the Bay Area into its natural watersheds and bay outlets so people living within them can work together to solve flooding and saltwater-intrusion problems.

258 Vietnam has lost nearly 60 percent . . . vulnerable to climate change: N. T. Hai et al., "Towards a More Robust Approach for the Restoration of Mangroves in Vietnam." *Annals of Forest Science* 77, no. 1 (2020): 1–18.

258 Without them, coasts can erode many yards each year: M. Spalding et al., *Mangroves for Coastal Defence: Guidelines for Coastal Managers & Policy Makers* (Wetlands International and the Nature Conservancy, 2014), 22.

258 Mangroves can also store almost three times as much carbon dixoide as rainforests: IUCN, "Issues Brief: Blue Carbon."

259 Mangroves and Markets: Nguyen Thi Bich Thuy, implementation project manager, Dutch Fund for Climate and Development in Vietnam, pers. comm., March 18–25, 2021.

259 The quantity varies a lot year to year, but the average is about 20 percent: Brenda Goeden, interview with author, December 4, 2020.

261 a football-field-sized piece of land . . . multiple causes: B. R. Couvillion et al., *Land Area Change in Coastal Louisiana (1932 to 2016)*, Scientific Investigations Map 3381, US Geological Survey, https://doi.org/10.3133/sim3381.

261 the amount of sediment needed in Louisiana is much greater than in San Francisco Bay: Sediment needed in Louisiana is about sixteen times more than what's need in San Francisco Bay. Sara Sneath, "Louisiana needs sand to rebuild its coast. Old oil and gas pipelines are blocking the way," Washington Post, August 5, 2021.

264 "I've seen pelicans, herons, shorebirds, and beautiful sunsets": Joshua Gies, pers. comm., October 28, 2020.

Chapter 10

265 It took forty-eight years . . . in just fourteen years: "New York's Sea Level Has Risen 9" since 1950," SeaLevelRise.org, https://sealevelrise.org/states/new-york/. The figure here is for the most likely forecast for the Battery from the US Army Corps of Engineers' sea-level change curve calculator, https://cwbi-app.sec.usace.army.mil/rccslc/slcc_calc.html.

266 six feet possible cited in a state report: *Recommendations to Improve the Strength and Resilience of the Empire State's Infrastructure* (NYS 2100 Commission), https://www.cakex.org/sites/default/files/documents/NYS2100.pdf.

266 $119 billion invested . . . or cope with storm runoff: Anne Barnard, "The $119 Billion Sea Wall That Could Defend New York . . . or Not," *New York Times*, January 17, 2020.

266 homeowners in Oakwood Beach . . . moved away from the coast: "Buyout & Acquisition Programs," New York Governor's Office of Storm Recovery, https://stormrecovery.ny.gov/housing/buyout-acquisition-programs.

266 Liz Koslov: Liz Koslov, interview with author, March 3, 2021.

267　By 2100, high tides will likely . . . worldwide: S. A. Kulp and B. H. Strauss, "New Elevation Data Triple Estimates of Global Vulnerability to Sea-Level Rise and Coastal Flooding," *Nature Communications* 10, no. 4844 (2019).

268　"The only way to win against water is not to fight": A. R. Siders, Miyuki Hino, and Katharine J. Mach, "The Case for Strategic and Managed Climate Retreat," *Science* 365, no. 6455 (2019): 761–63.

268　land loss in coastal Louisiana . . . between 2000 and 2010: The dire situation in Louisiana has led to just such planning. In 2019, the state published a blueprint for pulling people back from the coast and making space inland. See Louisiana's Strategic Adaptations for Future Environments, https://lasafe.la.gov/home/.

268　One in ten new homes . . . highest risk of flooding: Josh Halliday, "One in 10 New Homes in England Built on Land with High Flood Risk," *Guardian*, February 19, 2020, https://www.theguardian.com/environment/2020/feb/19/one-in-ten-new-homes-in-england-built-on-land-with-high-flood-risk.

269　In California . . . large residential subdivisions: Senate Bills 610 and 221.

269　A town in Utah . . . dwindling water supply: Jack Healy and Sophie Kasakove, "A Drought So Dire That a Utah Town Pulled the Plug on Growth," *New York Times*, July 20, 2021.

269　every dollar invested . . . avoided flood damages: Kris A. Johnson et al., "A Benefit–Cost Analysis of Floodplain Land Acquisition for US Flood Damage Reduction," *Nature Sustainability* 3, no. 1 (2020): 56–62.

269　The Army Corps of Engineers . . . stop defending against floods: Christopher Flavelle, "U.S. Flood Strategy Shifts to 'Unavoidable' Relocation of Entire Neighborhoods," *New York Times*, August 26, 2020.

269　Canada is already going down a similar path: Christopher Flavelle, "Canada Tries a Forceful Message for Flood Victims: Live Someplace Else," *New York Times*, September 10, 2019; Kimberley Molina, "'We have no choice': Flooded Gatineau Residents Mull Buyouts," *CBC News*, May 22, 2019.

269　Canada can implement such caps, while the United States cannot: A. R. Siders, interview with author, March 4, 2021.

270　the United States has $1.07 trillion . . . at risk of chronic flooding by 2100: *Underwater: Rising Seas, Chronic Floods, and the Implications for US Coastal Real Estate* (Union of Concerned Scientists, June 2018), https://www.ucsusa.org/sites/default/files/attach/2018/06/underwater-analysis-full-report.pdf.

270　California's state legislature proposed . . . Governor Newsom vetoed it: Nathan Rott, "California Has a New Idea for Homes at Risk from Rising Seas: Buy, Rent, Retreat," *All Things Considered*, NPR, March 21, 2021.

270　Alaska Native and Native American communities are already seeing flooding: Anna V. Smith, "Tribal Nations Demand Response to Climate Relocation," *High Country News*, April 1, 2020.

271　Nicholas Pinter: Nicholas Pinter, interview with author, July 17, 2020.

271　Pinter and Rees found some common threads: Nicholas Pinter and James C. Rees, "Assessing Managed Flood Retreat and Community Relocation in the Midwest USA," *Natural Hazards*, February 13, 2021.

272　$6 billion in levee improvements . . . to remove buildings from floodplains: David Conrad, *Higher Ground: A Report on Voluntary Property Buyouts in the Nation's Floodplains* (National Wildlife Federation, 1998), https://www.nwf.org/~/media/PDFs/Water/199807_HigherGround_Report.ashx.

272 the area's worst drought in at least five hundred years: B. I. Cook et al., "Spatiotem-
poral Drought Variability in the Mediterranean over the Last 900 Years," *Journal of
Geophysical Research: Atmospheres* 121 (2016): 2060–74.

273 failed crops . . . 1.5 million people off their land: Francesco Femia and Caitlin E.
Werrell, "Syria: Climate Change, Drought, and Social Unrest," Center for Climate
and Security, Briefer No. 11, February 29, 2012, https://climateandsecurity.files.
wordpress.com/2012/04/syria-climate-change-drought-and-social-unrest_
briefer-11.pdf.

273 Syria's devastating . . . weak governance: Erica Gies, "The Unseen Trigger behind
Human Tragedies," *TakePart*, October 12, 2016, http://www.takepart.com/feature
/2016/10/24/hidden-connections-climate-change.

273 severe drought from 2014 . . . forcing people off their land: Oliver-Leighton Barrett,
"Central America: Climate, Drought, Migration and the Border," Center for Climate and Security, April 17, 2019, https://climateandsecurity.org/2019/04/central-
america-climate-drought-migration-and-the-border/.

273 Human-caused climate change is a factor: A. Park Williams et al., "Large Contribution from Anthropogenic Warming to an Emerging North American Megadrought," *Science*, April 17, 2020.

274 Saudi Arabia has bought up . . . unsustainable water : Ian James and Rob O'Dell,
"Megafarms and Deeper Wells Are Draining the Water beneath Rural Arizona—
Quietly, Irreversibly," *Arizona Republic*, December 5, 2019, https://www.azcentral
.com/in-depth/news/local/arizona-environment/2019/12/05/unregulated-
pumping-arizona-groundwater-dry-wells/2425078001/.

274 1.8 million properties at risk of coastal flooding and erosion: UK Parliament,
Coastal Flooding and Erosion, and Adaptation to Climate Change: Interim Report
(UK Commons Select Committee on Environment, Food and Rural Affairs, 2019),
https://publications.parliament.uk/pa/cm201919/cmselect/cmenvfru/56/5604
.htm

274 it cannot win a war against the sea: *National Flood and Coastal Erosion Risk Management Strategy for England* (UK Environment Agency, 2020), https://assets
.publishing.service.gov.uk/government/uploads/system/uploads/attachment_
data/file/920944/023_15482_Environment_agency_digitalAW_Strategy.pdf.

275 Pete Hughes: Pete Hughes, interview with author, March 13, 2020.

275–76 Archaeologists found a submerged oak forest from Neolithic times . . . lined with
breeze blocks of chalk: Chichester and District Archaeology Society, "Medmerry
Beach Archaeology," June 17, 2020, YouTube video, 29 minutes, 41 seconds, https://
www.youtube.com/watch?v=fk9J263vXyo.

276 In modern times . . . erected where East Thorney once lay: Toru Higuchi et al.,
"Medmerry Realignment Scheme: Design and Construction of an Earth Embank-
ment on Soft Clay Foundation," in *From Sea to Shore—Meeting the Challenges of the
Sea* (ICE Publishing, 2014).

276 Every winter, workers would go out to the shore . . . annually: Ben McAlinden,
"Managed Realignment at Medmerry, Sussex," Institution of Civil Engineers, September 28, 2015, https://www.ice.org.uk/knowledge-and-resources/case-studies
/managed-realignment-at-medmerry-sussex.

276 "Work during winter storms . . . wasn't really working, to be honest": Pippa Lewis,
interview with author, April 8, 2020.

278 David Rusbridge: David Rusbridge, interview with author, November 27, 2020.

280 While it will continue to look for opportunities . . . to micromanage it: Pippa Lewis (UK Environment Agency, environmental project manager), email with the author, June 8, 2021.

281 Rowenna Baker: Rowenna Baker, interview with author, December 4, 2020.

282 "some of the worst weather for twenty years . . . working as planned": "Medmerry Coastal Flood Defence Scheme," UK Environment Agency, March 2014, https://www.gov.uk/government/publications/medmerry-coastal-flood-defence-scheme/medmerry-coastal-flood-defence-scheme.

282 making more space for water in . . . Portsmouth and Southampton : Richard Davies, "Flood Defences at Medmerry, UK," FloodList, November 15, 2013, http://floodlist.com/europe/united-kingdom/flood-defences-medmerry.

283 Elizabeth English: Elizabeth English, interview with author, December 8 and 10, 2020.

284 Many of the lifted homes were still lower than Katrina's eighteen-foot high-water-mark: John Simerman, "Lower 9th Ward Is Still Reeling from Hurricane Katrina's Damage 15 Years Later," NOLA.com, Aug. 29, 2020, https://www.nola.com/news/katrina/article_a192c350-ea0e-11ea-a863-2bc584f57987.html.

289 Climate scientists warn . . . we have already begun to suffer: "Only 11 Years Left to Prevent Irreversible Damage from Climate Change, Speakers Warn during General Assembly High-Level Meeting," United Nations, March 28, 2019, https://www.un.org/press/en/2019/ga12131.doc.htm.

290 Deep Adaptation: Jem Bendell, a professor of sustainability, environment and economics at the University of Cumbria (England), wrote a notable essay on this topic; see Benell, "To Criticise Deep Adaptation, Start Here," Open Democracy, August 31, 2020, https://www.opendemocracy.net/en/oureconomy/criticise-deep-adaptation-start-here/.